節慶菜

6 個中西節日 × 48 道美味佳餚

萬里編輯委員會 編著

萬里機構

編者話

每年與家人共聚的佳節不少，冬至、農曆新年、端午節、中秋節、聖誕節等，有想過在溫暖的家大宴親朋、為家人送上祝福嗎？

沒有做菜的頭緒？又不懂如何設計菜單？

本書介紹六款適合不同節日的菜單，給你做菜的靈感，例如：

傳統的做冬日子，親自炮製和味汁煮中蝦、士多啤梨生炒骨、菊花杞子浸紅鮋……傳統與創意融合，吃過後全家人團圓幸福；

在迎接新一年來臨的開年飯，好意頭的菜式是飯桌的主角，古法蒸龍躉（龍騰飛躍）、橙皮絲蜜汁金蠔（黃金滿屋）、合桃蝦鬆生菜包（包羅萬有）……吃下好運，祈求來年事事順景。

還有西方節日如聖誕節，也可精心炮製各式應節佳餚，歡欣度過。

快將全年對家人的祝福，透過豐盛的菜式傳給他及她，溫暖全家！

目錄

中秋節 Mid-Autumn Festival

■ 主菜 Main Dishes

■ 湯品 Soup

■ 小吃 Savoury Food

聖誕節 Christmas

■ 沙律 Salad

■ 主菜 Main Dishes

■ 甜品 Dessert

家庭聚餐 Family Gathering Meal

■ 前菜 Appetizers

■ 主菜 Main Dishes

■ 甜品 Dessert

團年飯、開年飯禮節事項

中國人過年過節最注重禮數及意頭，享用團年飯及開年飯時注意以下幾點，讓來年萬事順景、笑口常開、家宅平安！

添飯──添福添壽
農曆新年期間的團年飯及開年飯，「添飯」可以取個好彩頭，多添多吃，有「添福添壽」之意，福壽齊來。

好意頭菜式
中國人最重視好意頭，是以過年的菜式多以廣東話諧音取名，以示吉利，例如「魚」及「餘」同音，取其音「年年有餘」，表示有餘有剩、未來一年豐足之意；豬脷代表「大吉大利」；鮮蝦有「嘻哈大笑」之意，所以在團年飯及開年飯會特別準備魚、豬手、髮菜、蠔豉等，取個好意頭、好運連連。

多菜多餸──豐盛之意
每道菜的材料預備多些份量，讓飽餐之後碟內有餘有剩，代表豐盛的好意頭。

碗筷要輕放
傳統中國餐桌禮儀，動作要輕，移動餐具時不要發出刺耳的聲響；盡量小心不要打破飯碗，以免帶來厄運。

説吉利説話
農曆年間事事講求吉利，尤其吃團年飯、開年飯時，多説吉祥話語，席間保持笑容，予人精神奕奕之感。

揀好料竅訣

要煮出一道好餸，挑選上好的食材是首要條件，尤其在農曆年間的意頭好菜，怎少得生猛活跳的海鮮、肥美的鮮雞、美味的海味乾貨呢？

海產

※ 挑選游動、活動能力高、魚身帶光澤、表皮無破損的鮮魚。

※ 如選購原條冰鮮魚，必須察看以下特徵：魚鱗鮮明；魚鰓鮮紅；魚眼晶瑩、不帶灰暗；魚肉有彈性。

※ 被繩子綑綁的活蟹，難以得知其活動能力；但可留意以下幾點：眼睛要精靈多動；口吐泡沫；蟹鉗及蟹爪完整無脫落。

※ 鮮蝦以眼睛晶瑩突出；蝦身完整；蝦殼明亮有光澤；肉質有彈力的為佳。

※ 購買冰鮮蝦時，要注意蝦頭及蝦殼緊連不脫落；蝦頭不發黑；蝦殼鮮明、無黑點。

※ 見鮑魚互相黏附或吸附在玻璃上，輕按其肉立即收縮，都是新鮮活鮑魚，若無上述的特徵，建議不應購買。

雞隻

※ 活雞肉質富彈性、肉味濃郁、鮮味強。市面有售的芝麻雞，肉質嫩滑、骨軟，但味不濃。黃油雞皮下脂肪多，肉質細嫩、肉厚，雞皮滑溜。走地雞由於運動量足夠，雞肉結實、有嚼勁，肉味濃，皮下脂肪較少。

※ 冰鮮雞是將活雞宰殺後立即冰封及包裝，雞肉鮮明；但由於經冷藏，雞肉會鬆弛，肉質較軟。

海味乾貨

※ 乾鮑以每斤的「頭數」來計算，頭數愈少，鮑魚則愈大隻。選購時以鮑身肥美、底邊闊為佳。

※ 選購海參時，要檢查海參是否乾身、結實、肉厚，另外肉刺要完整；如海參含有水分則容易變質。浸發後的海參，肉質有彈性。

※ 花膠以乾身、色澤淡黃、呈半透明、無腥味的為佳。選購時放在燈光下，發現花膠色灰暗、無光澤，則表示在曬製時內部未完全乾透，積聚了水分，時間久了會發霉，不宜購買。

※ 乾瑤柱要購買個體完整、肉質結實、色澤金黃、表面乾爽、有鮮香味的為佳。日本出產的乾瑤柱香味濃郁、清甜，質量較高。

※ 乾冬菇經常是宴席的材料，以菇身完整飽滿、肉厚乾爽、紋理清晰、香味濃郁為佳。另外，可查看菇邊緊密有致、傘內底部呈淡黃色的褶紋。

準備宴席小秘訣

無論是過年宴會或是日常宴請友人回家飯聚,在預備美食、佈置裝飾及其他方面,每個細節都可費心思考,為賓客設想周到,令他們開懷歡聚!

設計菜單

※ 向來賓詢問有否食物過敏病史,盡量避免引起致敏的情況發生,較常引起致敏的食物如雞蛋、魚、蝦蟹或奶類等,可嘗試以其他食材代替。

※ 如席間有老人家或小朋友,建議魚肉去骨起肉;肉類切得較小塊,或以較軟腍的冬瓜、南瓜、帶子、蒸蛋白等作為餐單;另外,小朋友應避免進食用酒醃製的菜式,如醉雞等。

※ 預早清楚知道哪位賓客是茹素者,為他們安排特別的餐單菜式。

※ 菜式的味道不宜太濃重,緊記鹹淡相宜,品嚐濃味菜式後,配搭一款清新的食品,有一種洗滌味蕾的感覺。

※ 建議別偏重某類食材,海鮮、肉類、家禽、蔬菜、菇菌等食材排入你的菜單內,可豐富宴席的內容。

※ 家庭或聖誕聚會時,可嘗試設計幾道前菜或餐前小吃,在等候入席前作為聯誼交談時享用。

※ 如時間許可,安排飯後甜品,為宴席完結前留下一道甜美的回憶。

預先準備食材

※ 前菜或冷盤食物可預先製作妥當，冷吃的取出即可食用；熱吃的上菜前翻熱或炒煮即可。

※ 需要浸發後烹調的食材，如冬菇、猴頭菇、海參等，宜早一天浸發妥當，並冷藏備用。

※ 蝦蟹、魚等海鮮不宜太早沖洗或處理，以免鮮味流失。

※ 為了節省時間及令醃料更入味，需要長時間醃製的全雞或肉類，拌勻醃料後，宜放雪櫃待一晚，省卻宴席當天的烹調時間。

※ 自製的醬汁或調味料，適宜預早一兩天備好，烹調時更得心應手。

※ 燜煮的菜式建議宴席當天預早烹調妥當，開席前翻熱即可上桌，令煮食時更靈活。

※ 可使用多種烹調小家電協助，如焗爐、蒸鍋、氣炸鍋、真空煲等，可同一時間烹煮多款菜式，節省時間。

冬至飯

冬至是中國傳統文化的一個大節日，古時是祭祀的重要日子。

常聽説：「冬至大如年」，在中國人的社會裏，冬至這天全家人回家團圓吃飯，以豐盛的美食聚餐，俗稱「做冬」。香港人非常重視冬至，預備的菜式以雞、蝦、魚為主，飯後以湯圓作為甜品，象徵一家人團團圓圓，幸福和睦。

在不同的地方，冬至時有不同的飲食習俗，如：中國北方人宰羊、吃餃子；蘇州人吃餛飩；上海人吃湯圓；杭州人吃年糕；廣州人吃臘味糯米飯；韓國人則吃紅豆粥等。

菜譜

主菜

和味汁煮中蝦

薑蔥炒肉蟹

三杯醬海鮮鍋

XO 醬帶子炒鮮露筍

士多啤梨生炒骨

菊花杞子浸紅䱽

湯品

姬松茸黃耳煲雞湯

適合
8～10人
享用

甜品

酒釀冬蓉流黃湯圓

和味汁煮中蝦

Prawns in Ketchup Sauce

【材料】
新鮮中蝦 1 斤
青、紅辣椒（長型）各 1 隻
乾葱 4 粒

【調味料】
茄汁 4 湯匙
豆瓣醬 3 茶匙
糖 2 茶匙
水 4 湯匙

Ingredients

600 g fresh prawns
1 green chilli (with long-shaped)
1 red chilli (with long-shaped)
4 cloves shallots

Seasoning

4 tbsp ketchup
3 tsp spicy bean paste
2 tsp sugar
4 tbsp water

【做法】

1. 中蝦剪去鬚腳，在蝦背剦開少許，挑去腸，洗淨，抹乾水分，灑少許胡椒粉拌勻。
2. 青、紅辣椒去蒂、去籽，洗淨，切碎；乾葱去外衣，洗淨，切碎。
3. 中蝦撲上乾粟粉，放入油鑊煎至兩面金黃色，盛起。
4. 原鑊下乾葱及青、紅椒炒香，蝦回鑊，下調味料炒勻煮片刻即成。

Method

1. Trim the prawns. Cut along the back of prawns with the scissors, devein. Rinse and wipe dry. Sprinkle with ground white pepper and mix well.
2. Remove the stalks and seeds from the chillies. Rinse and finely chop. Remove the outer skin of onion. Rinse and finely chop.
3. Coat prawns with cornflour. Deep fry in a wok until both sides golden brown. Set aside.
4. Stir fry shallots and chillies in the same wok until fragrant. Put the prawns back in. Add the seasoning. Stir fry for a while and serve.

小技巧 Cooking tips

- 這款醬汁甜酸中帶微辣，可刺激味蕾，開胃生津。
- 中蝦挑去腸後，更乾淨衞生。
- The sauce has a sweet-sour taste with mild spicy, which stimulate our appetite.
- After the intestines are removed, the prawns are cleaner and hygienic.

薑葱炒肉蟹

Stir-fried Male Mud Crab with Ginger and Spring Onion

【材料】

肉蟹 2 隻
老薑 2 塊
葱 8 條
紹酒 1 湯匙
粟粉 4 茶匙

【調味料】

鹽 2 茶匙
糖半茶匙

【做法】

1. 肉蟹反轉背面，蟹腹朝上，用刀在蟹身中間斬入（勿斬破蟹蓋），揭開蟹蓋，去掉蟹鰓及沙囊，洗淨，瀝乾水分，斬件，拍鬆蟹鉗，灑入粟粉拌匀。

2. 葱去鬚頭，洗淨，切段，分成葱白及青葱；老薑洗淨，用刀拍鬆。

3. 燒熱鑊下油 3 湯匙，下老薑及葱白爆香，加入蟹件炒片刻，潷酒炒匀，下調味料及滾水 3/4 杯，加蓋焗煮約 6 分鐘，待汁液收少，下青葱段炒匀即成。

Ingredients

2 male mud crabs
2 mature ginger
8 sprigs spring onion
1 tbsp Shaoxing wine
4 tsp cornflour

Seasoning

2 tsp salt
1/2 tsp sugar

Method

1. Turn over mud crab with the abdomen facing up. Chop into the middle part without cutting all the way through the shell. Lift the shell off. Remove the gills and stomach. Rinse and drain. Chop into pieces. Lightly crack the claws. Sprinkle with the cornflour and mix well.

2. Remove the root of the spring onion. Rinse and cut into sections. Separate the white part from the green. Rinse the mature ginger. Crush to loosen the flesh.

3. Heat up a wok. Add 3 tbsp of oil. Stir-fry the ginger and white part of the spring onion until scented. Put in the crab and stir-fry for a moment. Sprinkle with the Shaoxing wine and stir-fry evenly. Pour in the seasoning and 3/4 cup of boiling water. Cook for about 6 minutes with a lid on. When the sauce reduces, put in the spring onion and stir-fry evenly to finish.

小技巧 Cooking tips

- 炒煮前才處理鮮蟹，可保留鮮甜的蟹香味。
- 新鮮肉蟹的特徵如下：眼睛及爪腳靈活運動；口吐泡沫；蟹身有重量；按其腹部有硬實飽脹的感覺；沒帶阿摩尼亞味。
- 老薑的薑味濃郁，毋須去皮，用刀拍鬆後，香味能徹底散發。
- Chopping the crab before stir-frying, it can keep the sweet taste of the crab.
- Without the smell of ammonia, fresh mud crab has got active movement in the eyes, claws and legs with bubbles spilling from the mouth. It carries some weight and the abdomen is firm and plump to the touch.
- Mature ginger has a strong flavour. Crushing the ginger with the skin on helps fully spread its fragrance.

三杯醬海鮮鍋

Seafood Hotpot in Three-cup Sauce

【材料】

帶子 4 兩
蝦 8 隻
鮮魷魚 1 隻
魚柳 4 兩
三色甜椒各半個
九層塔 2 棵
薑 8 片
乾葱 6 粒

【三杯醬】

麻油、紹酒、生抽各 3 湯匙
鎮江醋 2 湯匙
冰糖碎 1 湯匙

Ingredients

150 g scallops
8 prawns
1 fresh squid
150 g fish fillet
1/2 yellow bell pepper
1/2 green bell pepper
1/2 red bell pepper
2 sprigs Thai basil
8 slices ginger
6 cloves shallot

Three-cup sauce

3 tbsp sesame oil
3 tbsp Shaoxing wine
3 tbsp light soy sauce
2 tbsp Zhenjiang black vinegar
1 tbsp rock sugar (crushed)

【做法】

1. 帶子解凍，洗淨，抹上少許粟粉；蝦剪去鬚及腳，挑腸，洗淨；鮮魷魚劏好，劃十字花，切塊；魚柳洗淨，切塊。
2. 三色甜椒去籽，洗淨，切塊。
3. 燒熱瓦鍋下油 1 湯匙，加入帶子略煎，盛起。
4. 原鍋下薑片及乾葱炒香，加入鮮魷魚、魚塊及蝦拌勻，注入三杯醬煮滾，下三色甜椒及帶子煮至海鮮全熟，最後加入九層塔煮滾即成。

小技巧 Cooking tips

- 最後加入帶子炒煮，炒至剛熟即可，以免久煮令肉質過韌。
- 用瓦鍋烹調，注意火力不宜太大，以免容易燒焦，而且用瓦鍋煮出的菜式，香味特濃，惹味好吃！
- Put the scallops in at last and cook them until just done. Overcooking would make them tough and rubbery.
- Make sure you control the heat well with clay pot. Clay pot conducts heat very quickly and the food may burn easily if cooked over high heat. It tends to make the food more flavourful with a rustic charm.

Method

1. Thaw the scallops. Rinse well. Rub them with cornflour. Cut off the antennae and feet of the prawns. Devein and rinse well. Dress the squid and make light crisscross incision on the inside of the squid. Cut into pieces. Rinse the fish well and cut into chunks.
2. Seed the bell peppers. Rinse well and cut into pieces.
3. Heat a clay pot. Add 1 tbsp of oil. Put in the scallops and pan fry for a while. Set aside.
4. Stir fry ginger slices and shallot until fragrant in the same wok. Add squid, fish fillet and prawns. Stir well. Pour in the three-cup sauce and bring to the boil. Add bell peppers and scallops. Cook until all seafood is done. Add Thai basil at last. Bring to the boil and serve.

XO 醬帶子炒鮮露筍

Stir-fried Scallops and Asparaguses in XO Sauce

一家團圓節慶菜

【材料】

急凍帶子 8 兩
鮮露筍 8 兩
鮮百合 2 球
小粟米 10 條
XO 醬 3 茶匙
紹酒 1 湯匙
蒜肉 4 粒

【醃料】

胡椒粉少許
粟粉 3 茶匙

【調味料】

蠔油 1 湯匙
生抽 2.5 茶匙
糖 1 茶匙
粟粉 1.5 茶匙
水 3 湯匙

Ingredients

300 g frozen scallops
300 g asparaguses
2 fresh lily bulbs
10 baby sweet corns
3 tsp XO sauce
1 tbsp Shaoxing wine
4 cloves garlic

Marinade

ground white pepper
3 tsp cornflour

Seasoning

1 tbsp oyster sauce
2.5 tsp light soy sauce
1 tsp sugar
1.5 tsp cornflour
3 tbsp water

小技巧 Cooking tips

- 急凍帶子由冷凍室移放至下層，慢慢解凍即可；若時間急趕，可用少許鹽水浸至半軟身，或置於保鮮袋，浸於凍水內待至半軟。
- 鮮百合容易氧化，未使用前浸水內能保持潔白色澤。
- Transfer the frozen scallops from the freezer to a lower shelf of the refrigerator. Leave it them to thaw slowly until soft. If you're pressed for time, you may soak them in a little lightly salted water, or pack them in the ziplock bag and then soak in the water until half soft.
- Fresh lily bulb gets oxidized easily when exposed to air. Soaking them in water helps keep them white.

【做法】

1. 帶子解凍，洗淨，抹乾水分，下醃料拌匀。

2. 鮮百合切去焦黃部分，撕成瓣狀，洗淨，用水浸過面備用。

3. 鮮露筍只取翠嫩部分，洗淨，切斜段；小粟米切段，洗淨；鮮露筍及小粟米汆水，瀝乾水分。

4. 燒熱鑊下油2湯匙，先下帶子略煎，盛起。

5. 原鑊下蒜肉及 XO 醬炒香，下鮮露筍、小粟米及鮮百合炒匀，加入調味料（加入時不斷翻炒），最後下帶子，灒酒再炒片刻即成。

Method

1. Thaw the scallops. Rinse well and wipe dry. Add marinade and mix well.

2. Cut off the dark yellow base of the lily bulb. Tear into scales. Rinse well. Soak it completely in water. Set aside.

3. Cut off the fibrous end of the asparaguses. Rinse well. Cut into short lengths at an angle. Set aside. Cut the baby sweet corns into short lengths. Rinse well. Blanch the asparaguses and baby sweet corns in boiling water. Drain.

4. Heat a wok and add 2 tbsp of oil. Put in the scallops and pan fry for a while. Set aside.

5. Stir fry garlic and XO sauce until fragrant in the same wok. Put in the asparaguses, baby sweet corns and fresh lily bulbs. Toss well. Add seasoning while stirring continuously. Put in the scallops at last. Sizzle with wine and stir fry for a short while. Serve.

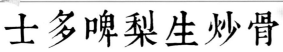

士多啤梨生炒骨

Sweet and Sour Pork Ribs in Strawberry Tamarind Sauce

【材料】

腩排 12 兩（斬塊）
士多啤梨 10 粒
青、黃甜椒各 1/2 個
羅望子 3/4 塊
乾粟粉 3/4 杯
油 2 湯匙

【醃料】

生抽 4 茶匙
紹酒 4 茶匙
胡椒粉少許

【調味料】

茄汁 3 湯匙
片糖碎 2.5 湯匙
鹽 3/4 茶匙

【芡汁】

粟粉 2 茶匙
水 6 湯匙
＊拌勻

Ingredients

450 g pork belly ribs (cut into cubes)
10 strawberries
1/2 green bell pepper
1/2 yellow bell pepper
3/4 slab dried tamarind
3/4 cup cornflour
2 tbsp oil

Marinade

4 tsp light soy sauce
4 tsp Shaoxing wine
ground white pepper

Seasoning

3 tbsp ketchup
2.5 tbsp raw cane sugar slab (crushed)
3/4 tsp salt

Thickening glaze

2 tsp cornflour
6 tbsp water
*mixed well

【做法】

1. 腩排洗淨，下醃料拌勻醃 1 小時。
2. 羅望子用滾水 1.5 量杯浸軟，過濾，隔渣備用。
3. 士多啤梨去蒂，洗淨，開邊；青、黃甜椒去籽，洗淨，切塊，用油炒熟備用。
4. 腩排均勻地蘸上乾粟粉，放入滾油內用慢火炸熟，再轉大火炸至金黃色，
   盛起，隔去油分。
5. 羅望子汁及調味料用慢火煮滾，煮至片糖溶化，埋芡，下油拌勻，最後
   加入腩排、甜椒及士多啤梨輕拌即成。

- 羅望子是印尼的調味料，於印尼雜貨店有售，帶天然的果酸香味。煮時要耐心地用慢火不時輕輕攪拌，以免醬料焦燶。
- 最後拌入少許油，令生炒骨的色澤油潤、光亮。
- Dried tamarind is Indonesian seasoning and is available in the Indonesian store. It is very fruity and tart. You should cook the sauce over low heat patiently. Stir occasionally to prevent it from burning.
- Stirring in a little oil at last makes the food shiny and glossy.

Method

1. Rinse the pork ribs. Add marinade and mix well. Leave them for 1 hour.
2. Soak dried tamarind in 1.5 cup of boiling water until soft. Strain. Set aside.
3. Remove the stems of the strawberries. Rinse and cut into halves. Set aside. Seed the bell peppers. Rinse and cut into pieces. Stir fry the bell peppers in a little oil until done. Set aside.
4. Coat the ribs evenly and lightly with cornflour. Deep fry in hot oil over low heat until done. Turn to high heat and fry until golden. Drain.
5. Boil the tamarind paste from step 2 and seasoning over low heat. Cook until the sugar dissolves. Stir in the thickening glaze. Stir in 1 tbsp of cooking oil. Put the ribs, bell peppers and strawberries back in. Toss well and serve.

菊花杞子浸紅䱽

Braised Snapper with Chrysanthemum and Wolfberry

【材料】

紅䱽魚 1 條（約 12 兩）

菊花 1/3 兩

杞子 2 湯匙

薑 2 片

芫茜 2 棵

鹽半茶匙

Ingredients

1 mangrove snapper (about 450 g)

13 g chrysanthemum

2 tbsp wolfberries

2 slices ginger

2 stalks coriander

1/2 tsp salt

【做法】

1. 菊花、杞子用熱水 1/3 杯略沖，隔去水分。

2. 紅鮋魚劏好，洗淨，抹乾水分。

3. 燒熱鑊下油 1 湯匙，下薑片拌香，加入熱水 2.5 杯煮滾，放入紅鮋魚、菊花，加蓋煮滾，轉小火浸煮約 12 分鐘，下杞子、鹽、芫茜煮滾，連湯盛起食用。

Method

1. Briefly rinse chrysanthemum and wolfberries with 1/3 cup of hot water. Drain.

2. Gut and rinse mangrove snapper, wipe dry.

3. Heat wok and add 1 tbsp of oil. Stir fry ginger until fragrant. Add 2.5 cups of hot water and bring to boil. Add mangrove snapper, chrysanthemum, cover the lid and bring to boil. Turn to low heat and cook for about 12 minutes. Add wolfberries, salt and coriander and bring to boil. Serve.

小技巧 Cooking tips

- 必須購買新鮮的紅鮋魚，抹淨魚肚內的瘀血，煮好的魚及湯才不會帶魚腥味。
- 用菊花及杞子浸魚，帶一陣清香之菊花香氣；而且由中醫角度來說，菊花及杞子有清肝明目的功能，改善視力模糊。
- Buy the fresh fish and sop up the blood in the gill. It helps reduce the fishy smell.
- Soaking the fish with chrysanthemum and wolfberry, you can taste the fish and soup with a scent of chrysanthemum. From Chinese medicine point of view, both chrysanthemum and wolfberry clear the Kidney and promote eye health, alleviate blurred vision.

姬松茸黃耳煲雞湯

Chicken Soup with Yellow Fungus and Agaricus Blazei Mushrooms

【材料】

姬松茸 2/3 兩

黃耳 2/3 兩

雪耳 2/3 兩

冰鮮雞 1 隻

瘦肉 6 兩

紅棗 10 粒（去核）

薑 4 片

Ingredients

25 g Agaricus Blazei mushrooms

25 g yellow fungus

25 g white fungus

1 chilled chicken

225 g lean pork

10 red dates (stoned)

4 slices ginger

【做法】

1. 黃耳用滾水加蓋浸 5 至 6 小時，洗淨，汆水。
2. 姬松茸、雪耳用水浸軟，摘去硬蒂，洗淨，汆水，瀝乾水分。
3. 雞去皮、去脂肪，洗淨，汆水，過冷河，瀝乾水分；瘦肉汆水。
4. 燒滾清水 18 杯，放入雞、瘦肉、黃耳、紅棗及薑片，用大火煲 20 分鐘，轉小火煲 1.5 小時，最後加入雪耳及姬松茸煲 45 分鐘，下鹽調味。

Method

1. Soak yellow fungus in boiling water for 5 to 6 hours and cover with the lid. Rinse and scald.
2. Soak Agaricus Blazei mushrooms and white fungus until soft. Remove the stems and rinse. Scald and drain.
3. Remove the skin and fat from the chicken. Rinse and scald. Rinse under the tap water for a while and drain. Scald the pork.
4. Bring 18 cups of water to the boil. Put in chicken, lean pork, yellow fungus, red dates and ginger. Boil over high heat for 20 minutes. Turn to low heat and simmer for 1.5 hours. Put in white fungus and Agaricus Blazei mushrooms and simmer for 45 minutes. Season with salt and serve.

小技巧 Cooking tips

- 去掉雞皮及脂肪才煲湯，減少油脂攝入量。
- 黃耳較硬，建議用滾水浸焗至軟身才煲湯。
- 姬松茸可提升身體的免疫功能，強身健體。
- Remove the skin and fat from the chicken before cooking the soup, it can reduce the fat intake.
- It is advisable to soak the yellow fungus in boiling water until soft before boiling the soup.
- Agaricus Blazei can boost the immune system and strengthen the body.

酒釀冬蓉流黃湯圓

Dumplings with Runny Yolk Filling in Fermented Glutinous Rice

【材料】

酒釀 12 湯匙

糯米粉 12 兩

冰糖 6 湯匙

【餡料】

熟鹹蛋黃 6 個（搓成蓉）

糖冬瓜 18 條（切碎）

花生蓉 6 湯匙

白芝麻 3 湯匙

牛油 6 茶匙

＊拌勻，冷藏 2 小時

Ingredients

12 tbsp fermented glutinous rice

450 g glutinous rice flour

6 tbsp rock sugar

Filling

6 cooked salted egg yolks (mashed)

18 strips candied winter melon (finely chopped)

6 tbsp finely chopped peanut

3 tbsp white sesame seeds

6 tsp butter

* mixed well and refrigerated for 2 hours

【做法】

1. 糯米粉與熱水 2 1/4 杯拌勻，搓成軟滑粉糰，搓至長形，分成小粒，按平後包入適量餡料。

2. 煮滾清水 6 杯，放入湯圓煮滾，加入冰糖及酒釀煮 7 分鐘，即可品嚐。

Method

1. Mix the glutinous rice flour with 2 1/4 cup of hot water. Knead into soft dough. Knead into a long strip. Divide into small dices. Flatten and wrap some filling in it.

2. Bring 6 cups of water to the boil. Put in the dumplings and bring to the boil. Add the rock sugar and fermented glutinous rice. Cook for 7 minutes. Serve.

小技巧 Cooking tips

- 餡料切碎或搓成蓉，可吃出幼滑的湯圓。
- 鹹香的流黃湯圓，伴濃郁的酒釀品嚐，味道非常匹配。
- You should chopped candied winter melon into fine pieces and mashed salted egg yolk for a smooth glutinous rice ball.
- The dumplings with runny yolk filling are salty and fragrant while the fermented glutinous rice is strong in taste. They are perfectly combined to give a wonderful sensation!

團年飯

團年飯（華北地區稱為年夜飯）在年廿九或年卅晚大除夕，一家人圍桌共吃，象徵全家人安好團圓，來年豐衣足食，是年末全家人最重要的一頓晚餐，共聚天倫，倍添家庭幸福。

對於年夜飯，中國傳統早有記載，古人認為年夜飯有驅邪逐疫的功效，全家人圍坐吃一頓豐盛的年夜飯後，大家身體健康，去除不吉利的事。

團年飯的菜式以好意頭為主，中國南方地區家庭必備頭尾完整的魚，諧音「餘」，有「年年有餘」之意；髮菜、蠔豉取廣東話諧音「發財好市」；蓮藕寓意聰明；湯圓寓意團團圓圓，是對來年的希望與期盼。

菜譜

主菜

背煎豉汁蟶子皇

XO 醬薑葱芹生蠔煲

黃金獅子頭

鮮滑白切雞

金華火腿蒸多寶魚

蠔汁三菇桃膠燴瓜脯

湯品

太子參猴頭菇豬腱螺片湯

適合
8~10人
享用

甜品

桂花珍珠紅豆沙

背煎豉汁蟶子皇

Fried Razor Clams in Black Bean Sauce

【材料】

蟶子皇 8 隻
蒜蓉 1 湯匙
豆豉蓉 1 茶匙
紅椒絲少許

【調味料】

生抽半湯匙
老抽 1 茶匙
糖 1/4 茶匙
＊拌勻

【做法】

1. 蟶子皇擦淨外殼，用刀在蟶子皇的中央位置直劖一刀，挑淨腸臟，洗淨，抹乾水分。
2. 平底鑊下油 2 湯匙，排上蟶子皇（外殼向下），用中火煎至蟶子肉收縮，上碟。
3. 原鑊加入蒜蓉及豆豉蓉爆香，下紅椒絲及調味料煮滾，澆在蟶子皇上，趁熱食用。

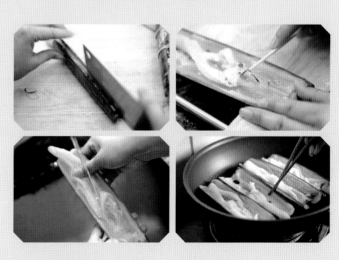

Ingredients

8 large razor clams
1 tbsp finely chopped garlic
1 tsp finely chopped fermented
black beans
shredded red chilli

Seasoning

1/2 tbsp light soy sauce
1 tsp dark soy sauce
1/4 tsp sugar
* mixed well

Method

1. Rub the razor clam shells clean. Gut the razor clams by straightly cutting into the middle of the body. Remove the intestine with the skewer. Rinse and wipe them dry.
2. Put 2 tbsp of oil in a pan. Line the razor clams on the pan with the shell down. Fry over medium heat until the razor clam meat shrinks. Put on a plate.
3. Stir-fry the garlic and fermented black beans in the same pan until scented. Add the red chilli and seasoning. Bring to the boil. Drizzle on the razor clams. Serve warm.

小技巧 Cooking tips

- 背煎法是將貝類的外殼排於鑊底煎煮，避免蟶子肉直接接觸明火，令肉質嫩滑，味道更鮮更濃。
- 見蟶子肉呈收縮狀即表示剛熟，必須立即上碟；否則蟶子肉久煮，肉質變韌。
- Put the shellfish in a pan with the shell down and the meat up. It can prevent the meat from direct heating, making it more smooth and delicious.
- When the razor clam meat is shrunken, it means it is just cooked and should be served immediately; Or the razor clam meat will become tough after being cooked for a long time.

XO 醬薑蔥芹生蠔煲

Simmered Oyster with Ginger and Spring Onion in XO Sauce

【材料】

生蠔 8 隻
薑 8 片
蔥 4 條
芹菜 2 棵
紅辣椒 1 隻（切絲）
XO 醬 1 湯匙
粟粉 2 湯匙

【調味料】

胡椒粉少許
紹酒 1 湯匙
生抽半湯匙

【芡汁】

粟粉 1.5 茶匙
水 2 湯匙
＊拌勻

Ingredients

8 fresh oysters
8 slices ginger
4 sprigs spring onion
2 stalks Chinese celery
1 red chilli (shredded)
1 tbsp XO sauce
2 tbsp cornflour

Seasoning

ground white pepper
1 tbsp Shaoxing wine
1/2 tbsp light soy sauce

Thickening glaze

1.5 tsp cornflour
2 tbsp water
*mixed well

【做法】

1. 生蠔用粟粉擦淨，用手檢走蠔殼碎片，洗淨，用滾水略灼，隔乾水分。
2. 葱切去根部，洗淨，切段；芹菜去根、去葉，洗淨，切段。
3. 砂鍋下油 2 湯匙，下薑片、葱段拌香，放入生蠔拌勻，下調味料、芡汁拌勻煮滾，加入 XO 醬、芹菜及紅椒絲拌勻，原鍋供食。

Method

1. Rub oysters with cornflour and pick out any oyster shells. Rinse oysters, scald and drain.
2. Cut off the roots from spring onion, rinse and cut into sections. Cut off the roots and leaves from Chinese celery, rinse and cut into sections.
3. Add 2 tbsp of oil in a clay pot. Stir fry ginger and spring onion until fragrant. Add oysters and mix well. Put in seasoning and thickening glaze, mix well and bring to boil. Mix in XO sauce, Chinese celery and red chilli. Serve in the pot.

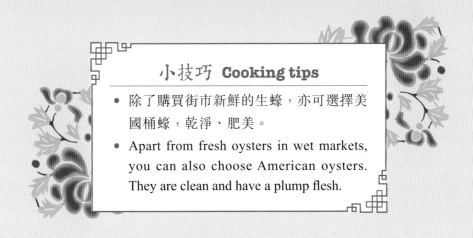

小技巧 Cooking tips

- 除了購買街市新鮮的生蠔，亦可選擇美國桶蠔，乾淨、肥美。
- Apart from fresh oysters in wet markets, you can also choose American oysters. They are clean and have a plump flesh.

黃金獅子頭

Simmered Salted Duck Egg Yolk
Meatballs

一家團圓節慶菜

【材料】

豬蹄肉 6 兩（攪碎或用免治豬肉）
鹹蛋 2 個
葱粒 1 湯匙
小棠菜 12 兩
薑 2 片
粟粉適量
麻油 2 茶匙

【醃料】

鹽半茶匙
生抽半湯匙
胡椒粉少許
粟粉 2 茶匙
水 3 湯匙

【調味料】

老抽 1 茶匙
生抽 1 湯匙
糖 1 茶匙

【芡汁】

粟粉 1 茶匙
水 2 湯匙
* 拌勻

Ingredients

225 g pork kunckle (ground or use minced meat)
2 salted duck eggs
1 tbsp finely chopped spring onion
450 g Shanghai white cabbage
2 slices ginger
cornflour
2 tsp sesame oil

Marinade

1/2 tsp salt
1/2 tbsp light soy sauce
ground white pepper
2 tsp cornflour
3 tbsp water

Seasoning

1 tsp dark soy sauce
1 tbsp light soy sauce
1 tsp sugar

Thickening glaze

1 tsp cornflour
2 tbsp water
*mixed well

小技巧 Cooking tips

- 建議勿用雙手搓肉丸，會令肉質結實，以湯匙協助令肉丸軟滑。
- It is recommended not to use hands to rub the meatballs; it will lead to tough meatballs.

【做法】

1. 鹹蛋黃焓熟，只取蛋黃，搓爛成蓉，備用。
2. 小棠菜一切為二，洗淨。
3. 免治豬肉與醃料由上至下拌勻至帶膠質，下鹹蛋黃蓉及葱粒拌勻，用湯匙搓成獅子頭，均勻蘸上粟粉，放入滾油內炸至表面金黃色，隔油。
4. 燒熱鑊下油 2 湯匙，下薑片炒香，下小棠菜炒軟，盛起。
5. 原鑊注入滾水 2 杯，放入獅子頭及調味料煮滾，用小火燜 30 分鐘，加入小棠菜續燜 5 分鐘，埋茨，下麻油拌勻。

Method

1. Hard-boil salted duck eggs. Take egg yolk only and finely crush. Set aside.
2. Cut Shanghai white cabbage into halves and rinse.
3. Mix pork with marinade and stir until sticky. Add grated egg yolk and spring onion and mix well. Shape a big meatballs with tablespoon. Evenly coat meatballs with cornflour. Deep fry in hot oil until golden brown. Drain.
4. Heat wok and add 2 tbsp of oil. Stir fry ginger until fragrant. Add Shanghai white cabbage and stir fry until soft. Remove.
5. Pour in 2 cups of boiling water in the same wok. Put in meatballs and seasoning and bring to boil. Turn to low heat and simmer for 30 minutes. Add Shanghai white cabbage and simmer for 5 minutes. Pour in thickening glaze and sesame oil. Mix well and serve.

鮮滑白切雞

Steamed Fresh Chicken with Spring Onion

【材料】
鮮活清遠雞 1 隻（約 3 斤）
蔥 6 條
薑汁半湯匙
粗鹽 1 湯匙

【蘸汁】
上等生抽 1 小碟

Ingredients
1 live Qing Yuan chicken (about 1.8 kg)
6 sprigs spring onion
1/2 tbsp ginger juice
1 tbsp coarse salt

Dipping sauce
1 small dish premium light soy sauce

【做法】
1. 清遠雞劏好，雞嘴及雞尾用粗鹽擦淨，再用水洗淨，瀝乾水分。

2. 雞腎刮淨內膜及脂肪，用粗鹽擦淨，再用水洗淨；雞肝去掉脂肪，用水洗淨。

3. 蔥去掉鬚頭，洗淨，切長段。

4. 雞內腔用粗鹽及薑汁塗勻，待半小時。

5. 將半份蔥段排在蒸碟底，放上鮮雞（雞胸向上），鋪上餘下之蔥段，與雞內臟一併隔水用大火蒸 25 分鐘，關火，加蓋焗 15 分鐘。

6. 預備半鍋盛有冰粒之凍開水，原隻雞放入冰水內浸凍，瀝乾水分，斬件上碟，伴以蘸汁供食。

小技巧 Cooking tips

- 清遠雞脂肪分佈平均，肉滑鮮美、外皮爽脆，而且飼養時自由走動，肉質結實，原隻蒸吃可嘗其原汁原味。
- 薑汁帶濃郁的香辛味，半湯匙份量塗抹雞內腔，已令鮮雞香氣四溢！
- 原隻雞蒸熟後，立即放入盛有冰粒之凍開水浸凍，因冷縮熱脹的關係，皮爽肉滑，滋味無窮！
- The free-range Qing Yuan chicken has a firm meat texture with even distribution of fat. Steamed whole chicken gives the original flavour with smooth meat and crispy skin.
- The pungent ginger juice makes the chicken very fragrant by just spreading 1/2 tbsp of the juice on the chicken cavity!
- Soak the steamed whole chicken into the water with ice cubes, the meat is firm and the skin is crunchy.

Method

1. Slaughter the chicken. Rub the beak and buttock with the coarse salt. Rinse and drain.
2. Scrape the membrane and fat off the chicken gizzard. Rub with the coarse salt and rinse. Remove fat from the liver. Rinse.
3. Cut away the root of the spring onion. Rinse and cut into long sections.
4. Spread the coarse salt and ginger juice on the chicken cavity. Rest for 1/2 hour.
5. Place 1/2 portion of spring onion on a plate. Put the chicken on top (breast upward). Arrange the rest of the spring onion on the chicken. Put the internal organs of the chicken on the side of the plate. Steam over high heat for 25 minutes. Turn off heat. Leave for 15 minutes with a lid on.
6. Prepare 1/2 pot of cold drinking water with ice cubes. Soak the whole chicken into the water to allow it to cool down. Drain. Chop into pieces. Serve with the dipping sauce.

金華火腿蒸多寶魚
Steamed Turbot with Jinhua Ham

【材料】

鮮活多寶魚 1 條（約 1 斤）
金華火腿 3 小片（切絲）
葱絲適量

Ingredients

1 live turbot (about 600 g)
3 small slices Jinhua ham (shredded)
shredded spring onion

【調味料】

生抽 2 茶匙

熟油 1 湯匙

Seasoning

2 tsp light soy sauce

1 tbsp cooked oil

【做法】

1. 多寶魚劏好，洗淨，抹乾水分，放蒸碟內，鋪上金華火腿絲。
2. 煮滾半鍋水，放入多寶魚，隔水大火蒸 6 至 8 分鐘（時間視乎魚的厚薄而定），傾掉半份蒸魚汁液，放上葱絲，澆上調味料即可。

Method

1. Gill the turbot, rinse and wipe dry. Put on a plate and then arrange the Jinhua ham on top.
2. Bring half pot of water to the boil. Steam the turbot over high heat for 6 to 8 minutes depending on the thickness of the fish. Discard half of the steamed fish sauce, put the spring onion on top and pour in the seasoning. Serve.

小技巧 Cooking tips

- 蒸魚時間視乎魚的厚薄大小而定，一般不多於 10 分鐘。
- 最後澆上熟油及生抽，令魚肉油潤及有鮮味。
- The time for steaming is generally less than 10 minutes, depending on the thickness and size of the fish.
- Splashing cooked oil and light soy sauce on the fish finally will make it moist, glossy and taste fresh.

蠔汁三菇桃膠燴瓜脯
Braised Chinese Marrow with Mushrooms and Peach Gum

【材料】

鮮冬菇 8 朵
靈芝菇 1 包
杏鮑菇 2 小朵
桃膠 2 湯匙
節瓜 3 個（約 1.5 斤）
薑 8 片

【調味料】

胡椒粉少許
麻油 2 茶匙
素蠔油 2 湯匙
粟粉 2 茶匙
蒸瓜汁 6 湯匙
＊拌勻

Ingredients

8 fresh black mushrooms
1 pack marmoreal mushroom
2 small oyster mushroom
2 tbsp dried peach gum
3 Chinese marrows (about 900 g)
8 slices ginger

Seasoning

ground white pepper
2 tsp sesame oil
2 tbsp vegetarian oyster sauce
2 tsp cornflour
6 tbsp juice from steaming Chinese marrows
*mixed well

【做法】

1. 桃膠用水浸 4 至 5 小時，挑去黑色污物，洗淨，放入滾水焓 3 分鐘，隔乾水分。
2. 節瓜刮淨外皮，直切，刮走瓜瓢，洗淨，切塊，排上蒸碟，灑入 1/4 茶匙海鹽，隔水中火蒸 12 分鐘，隔出汁液，保溫。
3. 鮮冬菇去蒂，洗淨，切粗條；杏鮑菇洗淨，切粗條；靈芝菇切去尾端，洗淨。
4. 燒熱鑊下油 3 湯匙，下薑片炒香，加入所有鮮菇、桃膠炒勻，下調味料煮 3 分鐘，淋上節瓜脯即成。

Method

1. Soak dried peach gum for 4-5 hours. Remove any black dirt. Rinse and scald for 3 minutes. Drain.
2. Peel off the skin from Chinese marrow, cut lengthwise in half. Rinse, cut into pieces and arrange on a plate. Add 1/4 tsp of sea salt and steam over medium heat for 12 minutes. Take the juice for the seasoning, keep the marrows warm.
3. Remove stalks from fresh black mushrooms, rinse and cut into thick strips; rinse oyster mushroom, cut into thick strips; cut off the root from marmoreal mushrooms and rinse.
4. Heat wok and add 3 tbsp of oil, stir-fry ginger until fragrant. Add all the mushrooms and peach gum and stir-fry well. Add seasoning and cook for 3 minutes. Transfer the mixture on top of Chinese marrows. Serve.

小技巧 Cooking tips

- 桃膠必須用水浸軟，以及挑去污物才烹調。
- 合掌瓜、青瓜、冬瓜等也非常合適燴煮，可吸收醬汁之精華。
- Dried peach gum should be soaked until soft and remove the dirt before cooking.
- Chayote, cucumber and winter melon are also suitable for this dish, as they can absorb the sauce well.

太子參猴頭菇豬腱螺片湯

Tai Zi Shen Soup with Conch and Monkey Head Mushroom

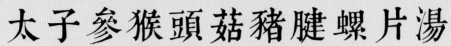

【材料】

太子參 2/3 兩
猴頭菇 2 兩
乾螺片 6 兩
豬腱 1.5 斤
甘筍 1 斤
薑 4 片

Ingredients

25 g Tai Zi Shen
75 g monkey head mushrooms
225 g g dried conch slices
900 g pork shin
600 g carrot
4 slices ginger

【做法】

1. 猴頭菇用水浸透（約1小時），剪去硬蒂，撕成小塊，洗淨，汆水，過冷河，擠乾水分。
2. 豬腱洗淨，切大塊，汆水，過冷河，洗淨。
3. 甘筍去皮，洗淨、切塊。
4. 太子參及乾螺片同洗淨。
5. 煲內注入清水 16 杯煲滾，放入全部材料用大火煲滾，續煲 15 分鐘，轉小火煲 1.5 小時，下少許鹽調味即成。

Method

1. Soak monkey head mushrooms thoroughly (about 1 hour). Cut off hard stems and tear into small pieces. Rinse, scald, rinse again and squeeze dry.
2. Rinse pork shin and cut into large pieces. Scald and rinse.
3. Peel carrot. Rinse and cut into pieces.
4. Rinse Tai Zi Shen and conch.
5. Bring 16 cups of water and bring to boil. Put in all ingredients and bring to boil over high heat. Boil for 15 minutes. Turn to low heat and simmer for 1.5 hours. Season with salt. Serve.

小技巧 Cooking tips

- 烹調前，猴頭菇要先浸軟及剪去硬蒂，撕成小塊才煲湯。
- Monkey head mushroom need to soak until soft, cut off hard stems and tear into small pieces before cooking the soup.

桂花珍珠紅豆沙

Red Bean Sweet Soup with Osmanthus and Tapioca Pearls

【材料】

紅豆 8 兩

台灣珍珠 6 湯匙

乾桂花 2 茶匙

陳皮 1 個

冰糖 4 湯匙

Ingredients

300 g red beans

6 tbsp Taiwanese tapioca pearls

2 tsp dried osmanthus

1 whole dried tangerine peel

4 tbsp rock sugar

【做法】

1. 陳皮用水浸軟，刮淨內瓢。
2. 紅豆洗淨，用水浸 2 小時，隔去水分。
3. 煲滾清水 4 杯，加入珍珠用小火煲 10 分鐘，關火焗 15 分鐘後，再煲 10 分鐘，關火再焗 15 分鐘，至珍珠完全軟透，盛起，過冷河，瀝乾水分備用。
4. 煲滾清水 12 杯，放入紅豆及陳皮，用大火煲滾，轉中小火煲 1 小時至紅豆軟脸，最後下冰糖、珍珠及桂花，煲至冰糖溶化即成。

Method

1. Soak dried tangerine peel in water to soften. Scrape off the pith.
2. Rinse the red beans. Soak in water for 2 hours. Strain.
3. Bring 4 cups of water to the boil. Add the tapioca pearls and simmer for 10 minutes over low heat. Turn off heat and leave for 15 minutes. Cook again for 10 minutes. Turn off heat and leave for 15 minutes until the tapioca pearls are completely soft. Remove and rinse in cold water. Drain and set aside.
4. Bring 12 cups of water to the boil. Put in red beans and dried tangerine peel. Bring to the boil over high heat. Turn to medium-low heat and cook for 1 hour until red beans are tender. Finally add rock sugar, tapioca pearls and osmanthus. Cook until the sugar dissolves. Serve hot.

小技巧 Cooking tips

- 煲至紅豆熟及大滾，見紅豆外殼浮於表面，即用隔篩濾去紅豆殼，令紅豆的澱粉質徹底熬出，入口幼滑。
- Cook the soup until it boils heavily and the red beans are cooked through. When the shells of red beans float on the soup, remove the shells with a sieve and let the starch inside the red beans fully release. The soup will then taste creamy.

開年飯

開年飯是農曆新年重要的一頓飯，迎接新一年來臨，祈求來年事事順利。

大年初一，傳統上向神明表示敬意，一般不殺生，吃齋菜取意吃掉一年的大災小禍。到了大年初二，大部分家庭準備開年，在新一年進行第一次祭祀的儀式。無論是家庭或做生意的人，大年初二會殺雞拜神、拜土地，大夥兒吃一頓豐富的飯菜，成為每年傳統的習俗。

正月初二是「頭禡」，祈求神明保佑全家人一年平安，故一家大小會聚首一堂吃飯，象徵來年團團圓圓、豐衣足食。開年飯的菜式以吉祥寓意為主，代表吃下好運，有魚有肉，取意「年年有餘」、「大魚大肉」，菜式豐富，讓全家人吃得開懷盡興。

菜譜

主菜

橙皮絲蜜汁金蠔（黃金滿屋）

三文魚籽帶子釀豆卜（連生貴子）

古法蒸龍躉（龍騰飛躍）

蝦籽蝦仁煮海參（心想事成）

蠔豉豬肉燜蓮藕（年年有餘）

合桃蝦鬆生菜包（包羅萬有）

湯品

淮杞排骨燉鮮鮑仔（萬事如意）

**適合
8～10人
享用**

甜品

八寶糖年糕（八福臨門）

橙皮絲蜜汁金蠔 (黃金滿屋)

Fried Oysters with
Orange Zest in Honey Sauce

一家團圓節慶菜

【材料】

新鮮生蠔 8 隻（約 12 兩）

橙皮絲 1 湯匙

蜜糖 2 茶匙

生抽 1 茶匙

紹酒 1 茶匙

粟粉 2 湯匙

【做法】

1. 生蠔用粟粉擦洗，用清水洗淨，放入
   滾水內略灼，盛起，過冷河，抹乾水
   分，風乾 3 至 4 小時 （見生蠔表皮乾
   爽即可）。

2. 平底鑊內燒熱油 2 湯匙，下生蠔煎至
   表面呈微黃色，灒紹酒，下橙皮絲及
   生抽拌勻，最後加入蜜糖拌至汁液濃
   稠即成。

Ingredients

8 fresh oysters (about 450 g)
1 tbsp shredded orange zest
2 tsp honey
1 tsp light soy sauce
1 tsp Shaoxing wine
2 tbsp cornflour

Method

1. Rub the oysters with cornflour. Rinse and slightly blanch. Remove and rinse in cold water. Wipe dry. Air-dry for 3 to 4 hours (when the surface of the oysters is dry, it is done).

2. Heat up 2 tbsp of oil in a pan. Fry the oysters until the surface is lightly brown. Sprinkle with the Shaoxing wine. Add the orange zest and light soy sauce. Mix well. Finally put in the honey and mix together until the sauce reduces. Serve.

小技巧 Cooking tips

- 生蠔汆水後，用廚房紙徹底抹乾水分，緊記盛於筲箕內置通風處吹至表面乾爽。

- Blanch the oysters and dry them thoroughly with kitchen paper. Then put them in a mesh strainer and set it in a well-ventilated place until the surface of the oysters is dry.

三文魚籽帶子釀豆卜 （連生貴子）

Scallop and Salmon Roe
Stuffed Tofu Puffs

【材料】

急凍大帶子 8 隻
豆卜 16 個
三文魚籽適量
粟粉適量

【調味料】

胡椒粉少許
生抽 3/4 湯匙
麻油 3 茶匙
* 拌勻

Ingredients

8 frozen large scallops
16 deep-fried tofu puffs
salmon roe
cornflour

Seasoning

ground white pepper
3/4 tbsp light soy sauce
3 tsp sesame oil
* mixed well

小技巧 Cooking tips

- 將帶子沾上粟粉，不易鬆脫，並盡量將帶子釀入豆卜內，帶子蒸熟後略收縮，賣相美觀。

- Dust the scallop with cornflour to help stick it to the tofu puff and try pushing the scallop inwards. The steamed scallop will shrink a little bit and look great.

【做法】

1. 急凍帶子解凍，略洗，抹乾水分，橫切兩厚片。

2. 用剪刀剪去豆卜的小塊，再按入豆卜內。

3. 厚帶子沾上粟粉，釀入豆卜內，排放蒸碟上，隔水大火蒸3分鐘，
   澆上調味料，再以三文魚籽裝飾食用。

Method

1. Defrost the scallops. Slightly rinse, wipe dry and coarsely slice.

2. Cut out a small piece from the top of the tofu puff with scissors and
   push it into the tofu puff.

3. Dust the scallops with cornflour and stuff into the tofu puffs. Arrange
   on a dish and steam over high heat for 3 minutes. Sprinkle with the
   seasoning and garnish with the salmon roe. Serve.

古法蒸龍躉 (龍騰飛躍)

Steamed Giant Grouper in Traditional Style

【材料】

新鮮沙巴龍躉 1 條（約 1 斤）

梅頭瘦肉 2 兩

冬菇 3 朵

荒茜 2 棵

熟油 1 湯匙

【醃料】

生抽 2 茶匙

老抽 1 茶匙

粟粉 1 茶匙

Ingredients

1 fresh Sabah giant grouper
(about 600 g)

75 g pork collar-butt

3 dried black mushrooms

2 stalks coriander

1 tbsp cooked oil

Marinade

2 tsp light soy sauce

1 tsp dark soy sauce

1 tsp cornflour

【做法】

1. 冬菇去蒂，用水浸 2 小時，擠乾水分，切絲；梅頭瘦肉洗淨，切絲。冬菇絲及肉絲用醃料拌勻，待半小時。
2. 芫茜去鬚根，洗淨，切段。
3. 沙巴龍躉劏好，洗淨，抹乾水分，放於蒸碟內，鋪上冬菇絲及肉絲。
4. 煮滾半鍋水，放入龍躉隔水大火蒸 12 分鐘，傾去半份蒸魚汁，放上芫茜，澆上熟油即成。

Method

1. Remove the stalks of the dried black mushrooms, soak in water for 2 hours, squeeze water out and cut into shreds. Rinse the pork collar-butt and cut into shreds. Mix the black mushrooms and pork with the marinade and rest for half an hour.
2. Remove the root of the coriander, rinse and cut into sections.
3. Gill the giant grouper, rinse and wipe dry. Put on a plate and lay the black mushrooms and pork on top.
4. Bring half pot of water to the boil. Steam the giant grouper over high heat for 12 minutes, discard half of the steamed fish sauce, put the coriander on top and then sprinkle with the cooked oil. Serve.

小技巧 Cooking tips

- 用廚房紙徹底抹去魚肚內之瘀血及去掉黑色薄膜，能夠辟走魚腥味。
- 最後撒上芫茜或葱，可增加魚的香氣。
- Removing the fishy smell, you should wipe away thoroughly the blood stasis and black membrane inside the fish belly with a kitchen paper towel.
- Coriander and spring onion are the popular spice among the Chinese cuisine and essential condiments for steamed fish. Spread them on the fish after it is steamed. It will give an additional sweet smell to the dish.

開年飯

蝦籽蝦仁煮海參 (心想事成)

Sea Cucumber with Shrimp and Shrimp Roe

Ingredients

450 g pre-soaked sea cucumber
225 g shelled shrimp
2 tsp shrimp roe
5 slices Jinhua ham
some spring onion sections
6 slices ginger
1 tbsp Shaoxing wine

Marinade

pinch of ground white pepper
2/3 tsp cornflour

Seasoning

2/3 tbsp oyster sauce
1 tsp golden caster sugar
1 cup water

Thickening glaze

2 tsp cornflour
4 tbsp water
*mixed well

【材料】
已浸發海參 12 兩
蝦仁 6 兩
蝦籽 2 茶匙
金華火腿 5 片
葱段適量
薑 6 片
紹酒 1 湯匙

【醃料】
胡椒粉少量
粟粉 2/3 茶匙

【調味料】
蠔油 2/3 湯匙
黃砂糖 1 茶匙
水 1 杯

【芡汁】
粟粉 2 茶匙
水 4 湯匙
＊拌勻

【做法】

1. 海參洗淨腸臟，切塊。
2. 煮滾清水半鍋，放入薑 3 片、蔥 2 條及海參汆水 2 分鐘，盛起海參，過冷河，瀝乾水分。
3. 蝦仁洗淨，抹乾水分，下醃料拌勻。
4. 燒熱鑊下油 2 湯匙，放入餘下薑片爆香，加入蝦仁炒片刻，盛起備用。
5. 原鑊下油 1 湯匙，放下海參，灒酒炒勻，下調味料加蓋煮 5 分鐘，加入蝦仁、金華火腿煮片刻，下芡汁、蔥段拌勻煮滾，上碟，最後灑上蝦籽享用。

Method

1. Rinse well sea cucumber. Cut into pieces.
2. Bring to boil half pot of water. Add 3 slices of ginger, 2 sprigs of spring onion and sea cucumber. Boil for 2 minutes. Dish up sea cucumber pieces and plunge into cold water. Drain.
3. Rinse shelled shrimps. Pat dry and combine with marinade ingredients.
4. Heat wok. Add 2 tbsp of oil. Stir-fry the remaining ginger until fragrant. Add shrimps and stir-fry briefly. Dish up.
5. Add 1 tbsp of oil to the same wok. Add sea cucumber. Sprinkle with wine and stir-fry until even. Add seasoning ingredients. Cover the lid and cook for 5 minutes. Add shrimps and Jinhua ham. Cook for a while. Add thickening glaze and spring onion sections. Stir-fry until even and bring to boil. Transfer to plate. Sprinkle with shrimp roe. Serve.

小技巧 Cooking tips

- 海參是高蛋白、低脂肪的食品，補腎，增強免疫力。若沒時間浸發，可選購市面已浸發的海參。
- 蝦籽如沒有香味或受潮影響，建議用白鑊烘香使用。
- Sea cucumber is a high-protein and low-fat food that is good for strengthening Kidney. It can help boost immunity and is a health enhancing food for all. To save time, get the pre-soaked and expanded sea cucumber from the market.
- It is recommended to stir fry the shrimp roe in the dry wok until fragrance if the shrimp roe is affected by the humid weather or no fragrance.

蠔豉豬肉燜蓮藕 (年年有餘)

Simmered Pork Shoulder, Dried Oysters and Lotus Roots

一家團圓節慶菜

【材料】

豬蹄肉 10 兩
蓮藕 12 兩
蠔豉 8 隻
薑 6 片
乾葱 3 粒（切片）
南乳 3/4 塊（搓爛）
麵豉醬半湯匙
芫茜 1 棵

【調味料】

老抽 2 茶匙
冰糖 2/3 湯匙
紹酒 1.5 湯匙

Ingredients

375 g pork shoulder
450 g lotus root
8 dried oysters
6 slices ginger
3 shallots (sliced)
3/4 cube red fermented bean curd (mashed)
1/2 tbsp soy bean paste
1 stalk coriander

Seasoning

2 tsp dark soy sauce
2/3 tbsp rock sugar
1.5 tbsp Shaoxing wine

小技巧 Cooking tips

- 燜煮的蓮藕要選購粗壯肥大的，咬入口易脫，而且粉糯好吃。
- 南乳是用大豆發酵的醬料，香氣濃，是煲仔菜的常用調味料，配搭肉類燜煮，惹味好吃。
- For simmering, choose strong and think lotus roots, they are soft and starchy.
- Red fermented bean curd made from fermented soybean. It is popular used seasoning in clay pot. With their strong taste and aroma, they go well with meat in clay pot.

【做法】

1. 蠔豉用熱水洗淨，備用。
2. 蓮藕洗淨污泥，切塊（如污泥太多，建議削去外皮）。
3. 豬蹄肉洗淨，切塊。
4. 砂鍋下油 2 湯匙，下薑片、乾葱爆香，放入蠔豉拌勻，下南乳、
   麵豉醬、豬蹄肉拌炒，加入熱水 3.5 杯、蓮藕、調味料煮滾，加
   蓋轉小火煮 1 小時，伴芫茜原鍋供食。

Method

1. Rinse dried oysters with hot water.
2. Rinse off any dirt or soil from lotus root and cut into pieces. Scrape off the peel if there is too much soil.
3. Rinse pork shoulder and cut into pieces.
4. Add 2 tbsp of oil in a clay pot. Stir fry ginger and shallots until fragrant. Mix in dried oyster and add red fermented bean curd, soy bean paste and pork shoulder. Stir fry well and add 3.5 cups of hot water, lotus roots and seasoning. Cover the lid and simmer over low heat for 1 hour. Top with coriander. Serve in the pot.

合桃蝦鬆生菜包 （包羅萬有）

Wrapped in Lettuce with Diced Shrimp and Walnut

【材料】

中蝦 10 兩
瘦肉（軟腍）4 兩
馬蹄 6 粒
合桃肉 3 兩
蒜肉 3 粒（拍鬆）
鹽半茶匙
紹酒半湯匙
沙律生菜 8-10 塊

【醃料】

生抽 2 茶匙
粟粉 1 茶匙

【調味料】

蠔油 1 湯匙
胡椒粉少許
麻油 2 茶匙

Ingredients

375 g medium shrimps
150 g tender lean pork
6 water chestnuts
113 g shelled walnuts
3 cloves skinned garlic (bashed)
1/2 tsp salt
1/2 tbsp Shaoxing wine
8-10 pieces lettuce for salad

Marinade

2 tsp light soy sauce
1 tsp cornflour

Seasoning

1 tbsp oyster sauce
ground white pepper
2 tsp sesame oil

小技巧 Cooking tips

- 炒蝦鬆可選體型細小的蝦，去殼、挑腸及切粒，方便烹調。

- Select small shrimps for dicing after shelling and removing the vein, it is more convenience for cooking.

【做法】

1. 合桃肉放入盛有半茶匙鹽的滾水內，焓 1 分鐘，盛起，抹乾水分；燒熱 4 湯匙油，用小火炒合桃肉至微黃，隔油備用。

2. 瘦肉洗淨，切粒，下醃料拌勻待半小時。

3. 中蝦去殼，挑腸，洗淨，抹乾水分，切粒。

4. 馬蹄去皮，洗淨，切粒；生菜洗淨，瀝乾水分，上碟。

5. 燒熱鑊下油 2 湯匙，下蒜肉爆香，蒜肉棄去，下瘦肉炒勻，加入蝦肉，灒酒炒勻，下馬蹄粒及調味料炒至蝦肉熟透，最後加入合桃略拌，上碟，以生菜包吃。

Method

1. Blanch the walnuts in boiling water with 1/2 tsp of salt for 1 minute. Wipe dry. Heat 4 tbsp of oil and stir-fry the walnuts over low heat until yellowish. Drain.

2. Rinse and dice the pork. Mix with the marinade and rest for half an hour.

3. Shell and devein the shrimps. Rinse, wipe dry and then cut into dices.

4. Shell the water chestnuts, rinse and dice. Rinse the lettuce, drain and then put on a plate.

5. Heat a wok and put in 2 tbsp of oil. Stir-fry the garlic until fragrant, discard the garlic, add the lean pork and stir-fry evenly. Put in the shrimps, sprinkle with the wine and give a good stir-fry. Add the water chestnuts and seasoning and stir-fry until the shrimps are cooked through. Finally mix in the walnuts. Dish up. Wrap in the lettuce to serve.

淮杞排骨燉鮮鮑仔 （萬事如意）

Double-steamed Abalone Soup with Huai Shan and Wolfberry

【材料】

鮮鮑仔 8 隻

唐排骨 1.5 斤

淮山半兩

杞子 2 湯匙

老薑 5 片

糯米酒 2 湯匙

海鹽 1 茶匙

Ingredients

8 fresh small abalones

900 g pork ribs

19 g Huai Shan

2 tbsp wolfberries

5 slices mature ginger

2 tbsp glutinous rice wine

1 tsp sea salt

【做法】

1. 淮山用溫水浸半小時，洗淨。
2. 唐排骨斬細件，洗淨，汆水，洗淨。
3. 鮮鮑仔原隻洗擦乾淨外殼，備用。
4. 將唐排骨、鮮鮑仔、薑片、淮山、糯米酒及滾水8杯放入燉盅內，加蓋，隔水大火燉半小時，轉小火燉1.5小時，加入杞子再燉半小時，下海鹽拌勻調味即成。

Method

1. Soak Huai Shan in warm water for 30 minutes. Rinse well.
2. Chop the pork ribs into small pieces. Rinse and scald the pork ribs. Rinse well.
3. Brush the abalone shell thoroughly and rinse well. Set aside.
4. Place pork ribs, abalones, ginger, Huai Shan, glutinous rice wine and 8 cups of boiling water in double-steamed pot. Cover with the lid. Stew over high heat for 30 minutes. Then turn to low heat and stew for 1.5 hours. Put in wolfberries and stew for 30 minutes. Season with salt. Serve.

小技巧 Cooking tips

- 必須用滾水同燉，切勿加入凍水，影響燉湯之功效。
- You must pour in the boiling water into the double-steamed pot, or it will be not as effective if you used cold water.

八寶糖年糕 (八福臨門)

Sweet Eight Treasure New Year Cake

【材料】

糯米粉 7 兩
粘米粉 1 兩
片糖 6 兩
蓮子、紅豆、眉豆、合桃、
　松子仁各 1 兩
紅棗 8 粒
杞子 1 湯匙
桂花 1 茶匙
油 1 茶匙

Ingredients

263 g glutinous rice flour
38 g rice flour
225 g raw cane sugar slab
38 g lotus seeds
38 g red beans
38 g black-eyed peas
38 g walnuts
38 g pine nuts
8 red dates
1 tbsp wolfberries
1 tsp dried osmanthus
1 tsp oil

【做法】

1. 蓮子去芯，用熱水加蓋浸 3 小時，隔去水分。
2. 眉豆、紅豆洗淨，用水浸 1 小時，隔去水分，隔水用中火蒸 1 小時至全熟，隔去水分備用。
3. 紅棗去核，洗淨，切碎；合桃洗淨，切碎；杞子洗淨，隔乾水分。
4. 糯米粉及粘米粉篩勻；片糖與水 1.5 杯煮成糖水，待涼。
5. 將糯米、沾米粉與片糖水拌成粉漿，加入蓮子、紅豆、眉豆、合桃、松子仁、紅棗及杞子拌勻，傾入已抹油之 8 吋糕盤內，灑入桂花鋪面。
6. 燒滾水，隔水大火蒸 1 小時，待涼，切塊享用。

Method

1. Core the lotus seeds. Soak them in hot water with a lid on for 3 hours. Strain.
2. Rinse the black-eyed peas and red beans. Soak in water for 1 hour. Strain. Steam over medium heat for 1 hour until they are fully cooked. Strain and set aside.
3. Core the red dates. Rinse and finely chopped. Rinse the walnuts and finely chopped. Rinse wolfberries and strain.
4. Sieve the glutinous rice flour and rice flour together. Cook the cane sugar slab with 1.5 cups of water into sweet soup. Let it cool down.
5. Mix the glutinous rice flour and rice flour with the sweet soup from step 4 into batter. Add the lotus seeds, red beans, black-eyed peas, walnuts, pine nuts, red dates and wolfberries. Mix well. Pour into a 8-inch greased cake container. Sprinkle dried osmanthus on top.
6. Bring water to the boil. Steam over high heat for 1 hour. Let it cool down. Cut into pieces and serve.

小技巧 Cooking tips

- 八寶糖年糕是溫州的食品，八寶料可自行決定，紅腰豆、雞心豆、百合或腰果等皆宜，只要預先浸洗或切碎即可。
- Sweet eight treasure new year cake is a delicacy in Wenzhou. You can choose any ingredients you like for the eight treasures. They could be red kidney beans, chickpeas, lily bulbs, or cashews. Just soak and rinse them, or chop them up beforehand.

中秋節

八月十五，中秋節，是中國人一個重要的傳統節日，寓意人月兩團圓。

在古代農耕之時，中秋節是農作物豐收的時節，中秋祭月有慶祝農耕秋收豐盛之意，感謝神恩，因此也稱為豐收節。

中秋節是一家人團圓的日子，人們會回家圍桌吃飯，以及感謝祖先庇佑。一般來說，中秋節吃月餅代表團團圓圓；但除了吃月餅，時令的農作物也是中秋節的應節食品。柚子是中秋節必備品之一，「柚」諧音「佑」，有保佑之意，吃果肉之外，柚皮也可做成特色菜餚。中秋吃芋頭有消災辟邪之意；桂花寓意富貴吉祥；蓮藕是團圓之意，此外還有秋季當造的栗子、石榴、螃蟹等。

菜譜

主菜

芋蓉腐皮卷

豆醬焗肉蟹

煎釀蓮藕夾

頭抽腩排燜柚皮

栗子冬菇燜乳鴿

芝麻肉桂芋絲球

湯品

桂花馬蹄魚蓉羹

適合
8~10人
享用

小吃

家鄉鹹湯圓

芋蓉腐皮卷

Deep-fried Taro Rolls in Beancurd Skin

【材料】

鮮腐皮 2 張
芋頭 1 個（1 斤 4 兩）
澄麵 4 湯匙
雞蛋 1 個（拂勻）

【調味料】

五香粉半茶匙
鹽半茶匙

【甜酸汁】

米醋 5 湯匙
茄汁 4 湯匙
片糖 3/4 塊（舂碎）
鹽 1/4 茶匙

Ingredients

2 sheets fresh bean curd skin
1 taro (750 g)
4 tbsp wheat starch
1 egg (whisked)

Seasoning

1/2 tsp five-spice powder
1/2 tsp salt

Sweet and sour sauce

5 tbsp rice vinegar
4 tbsp ketchup
3/4 slab brown sugar (crushed)
1/4 tsp salt

【做法】

1. 芋頭去皮，洗淨，切粗粒，隔水用大火蒸熟，趁熱用刀壓成芋蓉，加入澄麵及調味料，拌成芋泥。
2. 鮮腐皮剪去硬邊，用乾淨的濕毛巾略抹，抹上薄薄的芋泥（塗於腐皮 2/3 位置），捲成長條形，用蛋液黏緊。
3. 甜酸汁用慢火煮至片糖溶化，盛起備用。
4. 燒滾油，放入芋蓉卷，轉慢火炸至金黃色，盛起，瀝乾油分，切塊，伴甜酸汁蘸吃。

Method

1. Skin the taro. Rinse and coarsely dice. Steam over high heat until done. Mash with a knife while hot. Add the wheat starch and the seasoning. Mix well as taro puree.
2. Cut away the tough edge of the bean curd skin. Slightly wipe with a clean damp towel. Thinly spread the taro puree on 2/3 parts of the skin. Roll into a strip. Seal with the egg wash.
3. Cook the sweet and sour sauce over low heat until the brown sugar slab melts. Dish up and set aside.
4. Put the roll in hot oil. Then turn to low heat and deep-fry until golden brown. Dish up and drain. Cut into pieces. Serve with the sweet and sour sauce.

小技巧 Cooking tips

- 澄麵與芋蓉拌勻後，令芋蓉更黏稠。
- 於芋蓉卷兩端抹上生粉，以免芋泥容易溢出。
- 建議將整條芋蓉卷下油鍋炸透；如切塊炸會令芋蓉餡乾硬、欠口感。
- Wheat starch is to make the taro puree thick and gluey after mixing.
- Spread cornflour on both ends of the roll to seal in the taro puree.
- To avoid the filling turning hard and dry, it is better to deep-fry the whole roll first.

豆醬焗肉蟹
Mud Crabs with Puning Bean Sauce

【材料】
肉蟹 3 隻（約 2 斤）
潮州普寧豆醬 1 湯匙
薑 10 片
葱 8 條（切段）
粟粉 4 茶匙
紹酒適量

【調味料】
糖 1 茶匙
水 2/3 杯

Ingredients

3 mud crabs (about 1.2 kg)
1 tbsp Chaozhou Puning bean sauce
10 slices ginger
8 sprigs spring onion (sectioned)
4 tsp cornflour
Shaoxing wine

Seasoning

1 tsp sugar
2/3 cup water

【做法】

1. 肉蟹劏淨，處理妥當，洗淨，斬件；蟹鉗用刀拍裂，下粟粉拌勻。
2. 燒熱鑊，下油 6 湯匙，放入蟹件煎封，盛起。
3. 原鑊下薑片及半份葱段炒香，蟹件回鑊，灒酒拌勻，加入豆醬及調味料，加蓋，用中火焗煮 6 分鐘，最後放入餘下葱段拌勻即成。

Method

1. Remove the gills and guts of the crabs. Rinse and chop into pieces. Pound the claws with a knife. Mix well with the cornflour.
2. Heat a wok. Pour in 6 tbsp of oil. Fry to seal the crabs. Set aside.
3. Stir-fry the ginger and half of the spring onions in the same wok until fragrant. Put in the crabs. Sprinkle with the wine and mix well. Add the bean sauce and the seasoning. Cover with a lid and cook over medium heat for 6 minutes. Stir in the rest of the spring onions at last. Serve.

小技巧 Cooking tips

- 薑葱散發獨特的香氣，是起鑊炒蟹之最佳配料，令蟹件滲滿惹味的薑葱氣味。
- 煎封時，見蟹件轉成紅色即可，毋須完全熟透，因隨後需經焗煮步驟。
- 蟹鉗用刀拍裂，令香氣滲入蟹鉗內，而且食用時容易處理。
- With unique fragrance, a great amount of ginger and spring onion are the best ingredients for stir-frying crabs which give the appetizing smell of ginger and spring onion.
- Fry until the crabs turn red. It is not necessary to make them fully done because they will be cooked again later.
- Pounding the crab claws with a knife, it can let the aroma of the ingredients permeate the cracked claws. It also helps shell the claws easily when eating.

煎釀蓮藕夾
Fried Stuffed Lotus Root

一家團圓節慶菜

【材料】

中型蓮藕 2 節（約 1 斤至 1 斤 4 兩）
墨魚滑 6 兩
芫茜 3 棵
葱 2 條
粟粉 4 湯匙

【蘸汁】

陳醋 1 小碟

【做法】

1. 芫茜洗淨，取芫茜莖切碎；葱切去鬚根，洗淨，切粒。
2. 墨魚滑、芫茜碎及葱粒拌成餡料。
3. 蓮藕洗淨污泥，刮淨外皮，切掉兩端藕節，洗淨，切成圓片。
4. 蓮藕片抹上粟粉，釀入墨魚餡料 1 湯匙，蓋上另一塊蓮藕片，用力按實。
5. 燒熱平底鑊，下油 4 湯匙，排上蓮藕夾，加蓋，用中小火煎 5 分鐘，反轉再煎另一面呈金黃色，盛起，蘸陳醋享用。

小技巧 Cooking tips

- 釀入墨魚餡料並蓋上蓮藕片後，必須用力按實，以免煎熟後餡料分離。
- 加上鑊蓋用中火煎，以蒸氣熱力迅速將墨魚餡煎熟。
- After stuffing with the cuttlefish stuffing and cover with the lotus root slice, you should press them firmly, it can keep the stuffed lotus root intact after frying.
- By frying over medium heat with a lid on, the heat of the steam will make the cuttlefish stuffing done quickly.

Ingredients

2 section medium-sized lotus root (about 600 g to 750 g)
225 g cuttlefish paste
3 stalks coriander
2 sprig spring onion
4 tbsp cornflour

Dipping sauce

1 small plate black vinegar

Method

1. Rinse the coriander. Take the stems and chop up. Cut off the root of the spring onion. Rinse and dice.
2. Combine the cuttlefish paste, coriander stems and spring onion. Stir well as the stuffing.
3. Wash away the mud of the lotus root. Scrape the skin off. Cut away the joints on both ends. Rinse and cut into round slices.
4. Spread the cornflour on the lotus root slice. Fill in 1 tbsp of the cuttlefish stuffing. Cover with another lotus root slice. Press firmly.
5. Heat a pan. Add 4 tbsp of oil. Arrange the stuffed lotus root on the pan. Cover with a lid. Fry over medium-low heat for 5 minutes. Turn it over and fry the other side until golden. Serve with the dipping sauce.

頭抽腩排燜柚皮

Simmered Pomelo Pith and Pork Ribs

【材料】

腩排 10 兩（斬大塊）

柚皮 1 個

蝦米 3 湯匙

乾葱 3 粒（切片）

紹酒 3 茶匙

麻油 2 茶匙

【調味料】

頭抽 1.5 湯匙

老抽 1.5 茶匙

冰糖 2 湯匙

【芡汁】

粟粉 1.5 茶匙

水 3 湯匙

＊拌勻

Ingredients

375 g fatty pork ribs (chopped into big pieces)

1 pomelo pith

3 tbsp dried shrimps

3 cloves shallot (sliced)

3 tsp Shaoxing wine

2 tsp sesame oil

Seasoning

1.5 tbsp premium soy sauce

1.5 tsp dark soy sauce

2 tbsp rock sugar

Thickening glaze

1.5 tsp cornflour

3 tbsp water

*mixed well

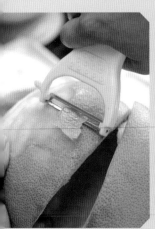

【做法】

1. 柚皮刨去皮，汆水數分鐘，啤水後壓乾水分，浸泡 1 至 2 天（期間換水），切塊備用。
2. 腩排洗淨，抹乾水分；蝦米洗淨。
3. 燒熱鑊下油 3 湯匙，下乾葱爆香，加入腩排及蝦米炒勻，潷紹酒炒勻，下調味料及水 3.5 杯煮滾，加入柚皮以慢火燜 50 分鐘，最後下芡汁及麻油煮滾即成。

Method

1. Peel out the zest of pomelo, then scald for a few minutes. Put it under the tap water. Drain and squeeze dry. Soak them in water for 1 to 2 days (change the water time after time). Cut the prepared pomelo pith into pieces. Set aside.
2. Rinse pork ribs and wipe dry. Rinse dried shrimps.
3. Heat wok and add 3 tbsp of oil. Stir fry shallots until fragrant. Add pork ribs and dried shrimps. Stir fry well. Sprinkle with Shaoxing wine and stir fry. Add seasoning and 3.5 cups of water. Bring to boil. Put in pomelo pith and turn to low heat. Simmer for 50 minutes. Pour in thickening glaze and sesame oil. Bring to boil and serve

小技巧 Cooking tips

- 柚皮經汆水、啤水及浸泡後，可去除苦澀味，能吸收肉汁之精華。
- After the process of scalding the pomelo pith, rinsing under the tap water and soaking in the water, the bitter taste of the pith will be removed. The pith will absorb the essence of the pork ribs after simmering.

栗子冬菇燜乳鴿

Simmered Pigeon with Chestnuts and Dried Black Mushroom

【材料】

乳鴿 2 隻
栗子 25 粒
乾冬菇 10 朵
乾葱 8 粒
薑 6 片
紹酒 1 湯匙

【醃料】

生抽 1 湯匙
紹酒 1 湯匙
粟粉 2 茶匙

【調味料】

蠔油 2 湯匙

Ingredients

2 baby pigeon
25 chestnuts
10 dried black mushrooms
8 cloves shallot
6 slices ginger
1 tbsp Shaoxing wine

Marinade

1 tbsp light soy sauce
1 tbsp Shaoxing wine
2 tsp cornflour

Seasoning

2 tbsp oyster sauce

【做法】

1. 冬菇去蒂，用水浸透，洗淨備用。
2. 栗子去殼，放入滾水內焓 3 分鐘，盛起，過冷河，去皮。
3. 乳鴿洗淨，每隻斬成 4 塊，下醃料醃半小時。
4. 燒熱鑊下油 3 湯匙，下栗子炒片刻，盛起。原鑊下薑片及乾葱爆香，加入乳鴿，潛酒炒勻，下冬菇及栗子拌炒，注入調味料及熱水 2.5 杯煮滾，轉小火燜約半小時，待栗子及乳鴿腍滑即可。

Method

1. Remove the stalks from black mushroom. Soak them in water until soft. Rinse well.
2. Remove the shells of chestnuts. Scald them in boiling water for 3 minutes. Remove and rinse under the tap water. Remove the skin.
3. Rinse baby pigeon. Chop each pigeon into quarters. Mix well with marinade. Leave for 30 minutes.
4. Heat wok and add 3 tbsp of oil. Stir fry chestnuts for a while. Set aside. Stir fry ginger and shallot in the same wok until fragrant. Put in pigeon and Shaoxing wine, stir fry well. Add black mushroom and chestnuts and stir fry. Pour in seasoning and 2.5 cups of hot water. Bring to boil. Turn to low heat and simmer for about 30 minutes until chestnuts and pigeon turn tender. Serve.

小技巧 Cooking tips

- 栗子先用油炒透，燜熟後能保持原整形狀。
- 入秋後是栗子的當造期，香甜味美，建議選外型圓渾、完整及帶外殼的。
- The shape of chestnut is set and fixed after stir frying, even stewed for a long time.
- The chestnut season is in autumn. It is a sweet taste ingredient. Pick those with round and plump shape with shell on.

芝麻肉桂芋絲球

Cinnamon Taro Balls with Sesame Seeds

【材料】

芋頭半斤
粘米粉 1 湯匙
炒香白芝麻適量

【調味料】

肉桂粉半茶匙
幼鹽半茶匙

【工具】

蘿蔔刨 1 個
小隔籬（打邊爐用）2 個

Ingredients

300 g taro
1 tbsp rice flour
toasted white sesame seeds

Seasoning

1/2 tsp ground cinnamon
1/2 tsp table salt

Tools

1 grater used for radish
2 small hot pot skimmers

【做法】

1. 芋頭去皮，洗淨，抹乾水分，刨成芋絲，加入調味料及粘米粉拌勻。

2. 小鍋內傾入半鍋油，將芋絲放在小隔籬內，放入滾油內用竹筷子略按實，炸片刻至形成球狀，脫離隔籬，再炸至金黃色，隔油上碟，灑下適量白芝麻即成。

Method

1. Peel the taro. Rinse and wipe it dry. Shred with a grater. Add the seasoning and rice flour. Mix well.

2. Pour oil to a small pot until it is half full. Put the taro in the skimmer. Put into the scorching oil. Press slightly with bamboo chopsticks. Deep-fry for a while until it turns to a ball shape. Free the ball from the skimmer. Deep-fry until golden. Drain and put on a plate. Sprinkle some white sesame seeds on top. Serve hot.

小技巧 Cooking tips

- 粘米粉的作用是黏緊芋絲，不容易散開來，份量毋須太多。
- 切掉芋頭一端，垂直放穩才去皮，能夠容易處理。
- Rice flour is used to stick the taro shreds together and so a small amount is enough.
- It is earlier to peel the taro skin when you cut away one end of taro first, then place the taro vertically and stably.

桂花馬蹄魚蓉羹

Fish Thick Soup with Water Chestnut and Osmanthus

【材料】

黃花魚 2 條（約 1.5 斤）
馬蹄 6 粒
雞蛋 3 個（拂勻）
乾桂花 2 茶匙
水 7 碗

【調味料】

魚露 1 湯匙
胡椒粉少許

【芡汁】

馬蹄粉 2.5 湯匙
水 5 湯匙
＊拌勻

Ingredients

2 yellow croaker (about 900 g)
6 water chestnuts
3 eggs (whisked)
2 tsp dried osmanthus
7 bowls water

Seasoning

1 tbsp fish sauce
ground white pepper

Thickening glaze

2.5 tbsp water chestnut powder
5 tbsp water
*mixed well

小技巧 Cooking tips

- 黃花魚肉質滑嫩，味道鮮美，非常適合烹調魚羹。
- 除了黃花魚，馬頭及青根也適合做成魚羹，這些魚的魚肉幼滑，魚味鮮美，只要小心挑淨魚骨即可。
- 馬蹄粉與水拌勻浸 15 分鐘，用篩子過濾後才埋芡，煮出來的湯羹更綿滑。
- With the smooth flesh, yellow croaker is suitable for making thick soup.
- Besides yellow croaker, yellow and silver horseheads are suitable because of their smooth texture and great taste. Make sure no bones end up in the soup!
- Mix the thickening glaze 15 minutes before use, and sift before adding to the soup. The thick soup is more smooth.

【做法】

1. 黃花魚劏好，洗淨，隔水蒸熟，待涼，去骨拆肉，保留魚肉。
2. 馬蹄去皮，洗淨，切碎。
3. 煮滾清水 7 碗，放入馬蹄碎及魚肉煮滾，下調味料煮滾，轉小火，下芡汁（一邊倒入一邊拌勻至合適的濃度），熄火，加入蛋漿拌勻，盛於碗內，灑下少量乾桂花享用。

Method

1. Prepare and gut yellow croaker and rinse. Steam the fish, let it cool, discard the bones and take only the meat.
2. Peel water chestnuts, rinse and finely chop.
3. Bring 7 bowls of water to boil. Add water chestnuts and yellow croaker and bring to boil. Add seasoning and bring to boil. Turn to low heat. Stir in thickening glaze until the desired thickness is achieved. Turn off heat and mix in egg wash. Transfer to the bowls. Sprinkle with dried osmanthus. Serve.

家鄉鹹湯圓
Savoury Rice Ball Soup

【湯圓材料】

糯米粉 6 兩

片糖 2/3 片（切大粒）

Ingredients for rice ball

225 g glutinous rice flour

2/3 piece brown slab sugar (cut into big cubes)

【材料】

鯪魚脊 4 條

鯪魚肉 6 兩

白蘿蔔 1 斤

乾葱 2 粒（切片）

蝦米 1.5 湯匙

冬菇 6 朵

芹菜 1 棵

鹽 1 茶匙

胡椒粉適量

Ingredients

4 strips backbone of dace

225 g dace paste

600 g radish

2 cloves shallot (sliced)

1.5 tbsp dried shrimps

6 dried black mushrooms

1 stalk Chinese celery

1 tsp salt

ground white pepper

小技巧 Cooking tips

- 鯪魚脊街市有售，價錢廉宜，煲出來的魚湯香甜美味。

- The soup made from dace backbone is sweet taste. The dace backbone is low in cost and available in the wet market.

【做法】

1. 鯪魚脊洗淨，抹乾水分，下鹽抹勻，放入油鑊煎至兩面金黃，注入熱水 8 杯煮滾，用中小火煲 1 小時，隔出鯪魚湯，備用。

2. 冬菇去蒂，用水浸 2 小時，擠乾水分，切絲；蝦米洗淨。

3. 白蘿蔔去皮，洗淨、切條；芹菜切去鬚根，摘去葉，洗淨，切粒。

4. 糯米粉加入適量清水，搓成軟滑粉糰，取一粒粉糰，按平，在中央放入一粒片糖，搓圓成湯圓，備用。

5. 燒熱鑊下油 1 湯匙，放入乾葱片、蝦米、冬菇爆香，注入鯪魚湯及熱水 (共 9 杯) 煮滾，下白蘿蔔煲 20 分鐘，鯪魚肉用湯匙逐少放入蘿蔔湯，煮至鯪魚肉全部浮起，加蓋保溫。

6. 煮滾適量清水，放入湯圓煮至浮起，再煮 3 分鐘，盛起。

7. 將鯪魚蘿蔔湯盛於碗內，加入適量胡椒粉、芹菜粒、湯圓食用。

Method

1. Rinse backbone of dace, pat dry and rub evenly with salt. Pan-fry in wok with a little oil until both sides turn golden. Add 8 cups of hot water. Bring to boil. Adjust to medium-low heat and simmer for 1 hour. Strain to set aside the dace soup.

2. Discard mushrooms stems. Soak mushrooms in water for 2 hours, squeeze dry and cut into thin shreds. Rinse dried shrimps.

3. Peel, rinse and cut radish into strips. Remove root and leaves of Chinese celery. Rinse and dice Chinese celery.

4. Add a little water to glutinous rice flour, knead to form smooth dough. Take some dough, flatten it and put a piece of brown slab sugar at the center. Knead to form rice ball. Set aside.

5. Heat wok. Add 1 tbsp of oil. Stir-fry shallot, dried shrimps and black mushroom until fragrant. Add dace soup and hot water (total 9 cups of liquid). Bring to boil. Add radish and cook for 20 minutes. Spoon dace paste little by little into radish soup. Cook until dace paste pieces float to the surface. Cover the lid and keep warm.

6. Bring some water to boil. Add rice balls and cook until rice balls float to the surface. Cook for another 3 minutes. Dish up.

7. Scoop dace radish soup into bowls. Add a little ground white pepper, Chinese celery and rice balls. Serve.

聖誕節

聖誕節，是基督教紀念耶穌降生的西方節日；時至今日，成為全家人歡聚、祝福的重要慶祝日子。

在平安夜或聖誕夜，大家準備豐盛的聖誕大餐與好友歡聚，共慶佳節，在不同的國家各有傳統的菜式，例如瑞典人以烤火腿為主；德國人的聖誕節主菜是烤鵝肉、薑餅；英國人的聖誕晚餐包括烤火雞、聖誕布甸、白蘭地奶油醬等等。

說起聖誕節的特色美食，令人聯想到聖誕火雞，原本火雞是美國人感恩節的主菜，後來成為聖誕的節慶食品，在西方人眼中，吃火雞有團圓之意。如覺得火雞的份量大，可改用烤雞代替，美味依然。除此之外，聖誕樹幹蛋糕、薑餅屋、聖誕布甸等也是受歡迎的食品。

菜譜

沙律
大蝦牛油果咯嗲

主菜
串燒一口牛
檸檬香草焗春雞
香草焗豬手
牛油蒜蓉黑椒煎蝦碌
芝士洋葱煙肉焗青口
帶子鮮蔬春卷

甜品
聖誕乾果蛋糕

適合
8~10人
享用

大蝦牛油果咯嗲
Prawn and Avocado Cocktail

【材料】
大蝦 8 隻
車厘茄數粒
西芹 1 條
青瓜粒 1 杯
牛油果 2 個
生菜葉數片
檸檬汁少許

【灼蝦料】
洋葱 1/3 個（切絲）
香葉 3 片

【調味料】
千島醬 6 湯匙

Ingredients
8 prawns
a few cherry tomatoes
1 piece celery
1 cup diced cucumber
2 avocadoes
a few slices lettuce
a dash of lemon juice

Ingredients for scalding prawns
1/3 onion (shredded)
3 bay leaves

Seasoning
6 tbsp thousand island salad dressing

【做法】

1. 大蝦去頭、挑去腸。
2. 燒滾水 1 杯，加入洋蔥絲、香葉煮滾，放入蝦灼熟，盛起，攤凍，去殼。
3. 青瓜粒用少許鹽醃片刻；車厘茄切半；西芹撕去筋，切粒。
4. 牛油果去皮、去核，切粒，灑下檸檬汁拌勻。
5. 盛器內先放上生菜葉，再放上青瓜粒、牛油果粒和西芹粒，舀上部分千島醬，再放上蝦和車厘茄，加入餘下的千島醬，雪藏片刻，可以品嚐這美味的沙律了。

Method

1. Remove the heads and dark dorsal vein from prawns.
2. Pour 1 cup of water in pot. Add shredded onion and bay leaves and bring to the boil. Put in prawns and scald until done. Remove and leave to cool. Remove the shells.
3. Marinate cucumber with a pinch of salt for a while. Cut cherry tomatoes in half. Tear the hard strings off celery and dice.
4. Peel and stone avocado. Dice. Sprinkle with lemon juice and mix well.
5. Place lettuce into container. Add cucumber, avocado and celery. Put some of the thousand island salad dressing onto the vegetables and fruit. Put in prawns and cherry tomatoes. Add the remaining thousand island salad dressing. Refrigerate for a while. Serve.

小技巧 Cooking tips

- 用洋蔥和香葉焓蝦，可去除腥味。
- 牛油果與檸檬汁調勻，可令牛油果不易變黑，賣相更美。
- Scalding prawns with onions and bay leaves can remove the unpleasant smell of prawns.
- Mixed with lemon juice, avocadoes will not turn brown and the salad will look nicer.

串燒一口牛
Skewered Beef Cubes

【材料】

急凍牛仔肉 3 條
三色甜椒各 1 個（小）
串燒竹籤 8 支

【醃料】

黑胡椒碎 1 茶匙
生抽 2 茶匙
喼汁 2 茶匙
粟粉 1 茶匙

【蘸汁】

黃芥末 3 茶匙
蜜糖 3 茶匙
*拌勻

Ingredients

3 strips frozen veal
1 each of three coloured bell peppers (small)
8 bamboo skewers

Marinade

1 tsp coarse ground black pepper
2 tsp light soy sauce
2 tsp Worcestershire sauce
1 tsp cornflour

Dipping sauce

3 tsp mustard
3 tsp honey
*mixed well

【做法】

1. 牛仔肉解凍，抹乾血水，切成一口小塊，下醃料拌勻。
2. 三色甜椒去蒂、去籽，洗淨，切成一口小塊。
3. 三色甜椒、牛仔肉用竹籤相間串好。
4. 燒熱油鑊放下油 1 湯匙，排入牛肉串燒，用大火每邊各煎 1 分鐘，上碟，塗抹蘸汁或伴吃。

Method

1. Thaw veal. Pat dry and cut into bite-sized pieces. Combine with marinade ingredients.
2. Remove stem and seeds of bell peppers. Rinse and cut bell peppers into bite-sized pieces.
3. Alternately thread bell pepper and veal pieces on bamboo skewers.
4. Heat pan. Add 1 tbsp of oil. Arrange veal skewers in pan. Pan-fry each side over high heat for 1 minute. Spread over the dipping sauce or dip as a sauce when serving.

檸檬香草焗春雞

Roasted Spring Chicken with Lemon and Thyme

【材料】
急凍春雞 2 隻（約 1.6 公斤）
檸檬 2 個
蒜肉 4 粒
原個蒜肉 2 個（連皮）
百里香 8 棵
甘筍 2 條
馬鈴薯 4 個（小）
橄欖油 5 湯匙

【醃料】
黑胡椒碎 2 茶匙
粗鹽 2 茶匙
紅酒 3 湯匙

【做法】
1. 急凍春雞解凍，洗淨，抹乾水分，用醃料將雞內外抹勻，放雪櫃醃一晚。
2. 甘筍及馬鈴薯去皮，洗淨，切塊。
3. 將蒜肉、百里香 2 棵、檸檬填入雞腔內，用竹籤串好，掃上橄欖油，放在網架上，備用。
4. 焗盤內放入甘筍、馬鈴薯、原個蒜肉、百里香，淋入餘下之橄欖油，放入已預熱 180℃ 之焗爐下層，春雞放在中層，焗 1 小時 15 分鐘，再將焗爐加至 200℃ 焗約 15 分鐘，待春雞外皮呈金黃即成，取出春雞待 10 分鐘，可伴焗蔬菜享用。

小技巧 Cooking tips

- 將蒜肉、百里香及檸檬塞進雞腔，令香味經烤焗過程滲入雞肉，美味好吃。
- 甘筍、馬鈴薯、蒜肉及百里香等放於春雞下烤，雞汁精華令蔬菜加添味道。
- Put garlic cloves, thyme and lemon inside the chicken, the flavour and aroma can still penetrate into the chicken through baking.
- Put the baking tray with carrot, potato, whole garlic and thyme in the lower part inside oven, the essence of chicken come from the baking process makes the vegetables tasty.

Ingredients

2 frozen spring chicken
  (about 1.6 kg)
2 lemons
4 cloves skinned garlic
2 whole garlic (with skin on)
8 sprigs thyme
2 carrots
4 small potatoes
5 tbsp olive oil

Marinade

2 tsp chopped black pepper
2 tsp coarse salt
3 tbsp red wine

Method

1. Defrost the spring chicken, rinse and wipe dry. Rub the marinade through outside and inside of the chicken. Put in the fridge overnight.
2. Peel carrot and potato, rinse and cut into pieces.
3. Put skinned garlic, 2 sprigs of thyme and lemon inside each chicken. Skew the chicken up and spread with olive oil. Set aside on baking rack.
4. Add carrot, potato, whole garlic and remaining thyme in baking tray. Spread over the olive oil. Preheat oven at 180°C, put the baking tray in the lower part inside oven and chicken in the middle. Bake for 1 hour 15 minutes. Turn to 200°C and bake for 15 minutes, until golden brown. Remove the chicken and let it sit for 10 minutes. Serve with the vegetables.

香草焗豬手

Baked Pork Knuckle with Herbs

【材料】

包裝鹹豬手 1 隻（大隻，未調味）
薑 4 片
黑椒碎 2 茶匙
混合乾香草適量

Ingredients

1 packed salt-cured pork knuckle (large-sized, not spiced)
4 slices ginger
2 tsp chopped black pepper
mixed dried herbs

小技巧 Cooking tips

- 若發熱線太貼近豬手，建議蓋上錫紙以免焦燶。
- 鹹豬手不要斬件烤焗，以免肉汁流失。
- 可減省黑椒碎及混合香草的份量，可吃出濃濃的肉香味。
- As the heating wire of the oven is near, it is better to cover the pork knuckle with aluminum foil to avoid scorching.
- It is not recommended to chop the pork knuckle into pieces before baking, or it makes the meat too dry.
- Reduce the quantity of chopped black pepper and mixed herbs to avoid overpowering. You can serve the meat full of meat flavour.

【做法】

1. 鹹豬手略洗，備用。
2. 燒滾水，放入薑片及鹹豬手煮滾，轉中小火煲 1 小時，盛起豬手待片刻。用黑椒碎抹勻豬手，灑入適量混合乾香草。
3. 將已醃味的豬手放在焗盤內，放入已預熱之焗爐，用 180℃ 上下火焗約 40 分鐘，待豬手表皮呈金黃色及散發陣陣烤肉香味，取出，待 15 分鐘，切塊享用。

Method

1. Slightly rinse the pork knuckle. Set aside.
2. Bring water to the boil. Put in the ginger and pork knuckle. Bring to the boil. Turn to low-medium heat and cook for 1 hour. Dish up. Leave for a while. Rub the pork knuckle with the chopped black pepper. Sprinkle with some mixed dried herbs.
3. Put the spiced pork knuckle into a baking tray. Bake in a preheated oven with both upper and lower heat at 180°C for about 40 minutes. When the pork knuckle gives a golden outside with a roast meat fragrance, take out and leave for 15 minutes. Cut into pieces and serve.

牛油蒜蓉黑椒煎蝦碌

Fried Prawns with Butter, Garlic and Black Pepper

【材料】
新鮮中蝦 1 斤
蒜蓉 2 湯匙
牛油 3 茶匙
紹酒 1 湯匙
粟粉 3 茶匙

【調味料】
幼海鹽 1 茶匙
黑胡椒碎 1 茶匙

Ingredients

600 g fresh prawns
2 tbsp finely chopped garlic
3 tsp butter
1 tbsp Shaoxing wine
3 tsp cornflour

Seasoning

1 tsp fine sea salt
1 tsp chopped black pepper

【做法】

1. 中蝦剪去蝦鬚及蝦腳，挑腸，洗淨，每隻鮮蝦切成兩段，抹乾水分，加入粟粉拌勻。

2. 燒熱鑊下油 3 湯匙，放入中蝦煎至轉成紅色，灒酒炒勻，下蒜蓉及調味料炒香，最後加入牛油炒至汁液收乾即成。

Method

1. Cut away the tentacles and legs of the prawns with a pair of scissors. Devein and rinse. Cut each prawn into halves. Wipe them dry. Add the cornflour and mix well.

2. Heat up a wok. Add 3 tbsp of oil. Fry the prawns until they turn red. Sprinkle with the Shaoxing wine and stir-fry evenly. Add the garlic and seasoning. Stir-fry until fragrant. Add the butter and stir-fry until the sauce dries. Serve.

小技巧 Cooking tips

- 最後加入牛油拌炒，令蝦肉帶濃濃的牛油香氣，享用時牛油味濃厚，惹味好吃！
- 毋須剝掉蝦殼，免蝦肉直接受熱，快炒可保持蝦肉嫩滑。
- Adding the butter in the final step, it gives the prawns a strong butter flavour, which is sensational!
- It is recommended not to remove the shell before cooking, you can try the smooth prawn flesh.

芝士洋蔥煙肉焗青口
Baked Mussels with Cheese, Onion and Bacon

【材料】

急凍半殼青口 10 隻
洋蔥 1/4 個
煙肉 1 片
芝士 2 片
鮮奶 1/3 杯
牛油 1/2 湯匙

【茨汁】

粟粉 1 茶匙
水 2 湯匙
＊拌勻

Ingredients

10 frozen half shell mussels
1/4 onion
1 slice bacon
2 slices cheese
1/3 cup milk
1/2 tbsp butter

Thickening glaze

1 tsp cornflour
2 tbsp water
＊mixed well

【做法】

1. 青口解凍，放入滾水略灼，盛起，瀝乾水分。
2. 洋蔥去外衣，洗淨、切絲；煙肉切絲，用少許油炒香洋蔥及煙肉備用。
3. 芝士撕成小塊，與鮮奶用小火煮至芝士溶化，加入牛油及芡汁煮滾成芝士汁。
4. 於青口面放上適量洋蔥煙肉，澆上芝士汁，排於焗盤上。放入已預熱之焗爐，用 180℃ 上下火焗至表面金黃（約 10 分鐘）即成。

Method

1. Defrost the mussels. Slightly blanch in the boiling water. Remove and drain.
2. Remove the outer skin of onion. Rinse and cut into shreds. Shred the bacon. Stir-fry the onion and bacon with a little oil until aromatic. Set aside.
3. Tear the cheese into small pieces. Cook with the milk over low heat until the cheese melts. Add the butter and thickening glaze. Bring to the boil as cheese sauce.
4. Put some onion and bacon on top of the mussels. Sprinkle with the cheese sauce. Lay on a baking tray. Put into a preheated oven. Bake at 180°C with both upper and lower heat until the surface is golden (about 10 minutes). Serve.

小技巧 Cooking tips

- 如芝士汁煮得很濃稠，可加入少許鮮奶略煮至稀；煮芝士汁時，最後緊記將芡汁逐少加入，見芝士汁濃稠程度適中即可。

- Add a little milk to cook the sauce thin if the cheese is thicken. When making the cheese sauce, remember to add the thickening glaze little by little. It is done when the sauce has a mild consistency.

帶子鮮蔬春卷

Spring Rolls Stuffed with Scallops and Greens

【材料】

急凍帶子 8 個
椰菜半斤
甘筍 2 兩
葱 2 棵
雞蛋 1 個 （拂勻）
急凍春卷皮 12 塊

【調味料】

蠔油 1 湯匙
胡椒粉少許

【蘸汁】

喼汁 1 小碟

Ingredients

8 frozen scallops
300 g cabbage
75 g carrot
2 sprigs spring onion
1 egg (whisked)
12 sheets frozen spring roll skin

Seasoning

1 tbsp oyster sauce
ground white pepper

Dipping sauce

a small plate Worcestershire sauce

小技巧 Cooking tips

- 炒蔬菜時以蠔油代替鹽調味，以免蔬菜釋出水分，令春卷皮濕潤。
- 餡料炒好後盛於不銹鋼碟內，座於凍水或冰水一會可加速冷卻。
- Season the vegetable filling with oyster sauce instead of salt. It avoids the vegetables releasing water, making the spring roll skin moist and sticky.
- To speed up cool down, put the vegetable filling on a stainless steel plate and set it on cold water or iced water.

【做法】

1. 帶子解凍，洗淨，抹乾水分，切粗條。

2. 甘筍去皮，洗淨，切絲；椰菜洗淨，切絲；蔥去鬚根，洗淨，切段。

3. 燒熱鑊下油 1 湯匙，下椰菜及甘筍炒至椰菜軟身，加入帶子、蔥段及調味料略拌，盛起待涼。

4. 春卷皮解凍，用濕毛巾蓋好，備用。

5. 每塊春卷皮鋪上適量餡料，在春卷皮邊緣塗上蛋漿，向上覆蓋 1/3 位置，兩邊向內摺入，捲成春卷狀。

6. 燒熱油，放入春卷炸至金黃色，隔油，切短度，伴喼汁吃。

Method

1. Defrost the scallops. Rinse and wipe dry. Cut into coarse strips.

2. Peel and rinse the carrot. Cut into shreds. Rinse the cabbage. Cut into shreds. Remove the root of the spring onion. Rinse and cut into sections.

3. Heat up a wok. Add 1 tbsp of oil. Stir-fry the cabbage and carrot until the cabbage is soft. Add the scallops, spring onion and seasoning. Slightly mix up. Leave to cool down.

4. Defrost the spring roll skin. Cover with a damp towel.

5. Lay some filling on each spring roll skin. Brush the egg wash along the edge of the skin. Fold up to 1/3 part of the skin. Fold both sides inwards. Roll up as a spring roll.

6. Heat up some oil. Deep-fry the spring rolls until golden. Drain and cut into short sections. Serve with the Worcestershire sauce.

聖誕乾果蛋糕
Christmas Fruit Cake

【材料】

黑加侖子乾 50 克

提子乾 80 克

杏脯乾 60 克

金提子乾 30 克

雜果皮 80 克

車厘子 30 克

榛子果仁 60 克（切碎）

冧酒 50 毫升

麵粉 100 克

發粉 1/3 茶匙

牛油 90 克

黃砂糖 80 克

雞蛋 2 個

肉桂粉、豆蔻粉各 1/4 茶匙

【餅面】

冧酒 20 毫升

Ingredients

50 g dried blackcurrants

80 g dried raisins

60 g dried apricots

30 g golden sultana

80 g mixed chopped peel

30 g glace cherries

60 g hazelnuts (shelled and chopped)

50 ml rum

100 g flour

1/3 tsp baking powder

90 g butter

80 g brown sugar

2 eggs

1/4 tsp ground cinnamon

1/4 tsp ground cardamom

Topping

20 ml rum

小技巧 Cooking tips

- 最後蓋上錫紙，可保存冧酒的香氣，而且以免蛋糕太乾，影響口感。

- Cover the cakes in foil at last, so that they would keep the rum flavour and would not be too dry.

【做法】

1. 所有乾果切碎,與榛子果仁碎用冧酒浸一晚(或更長時間)。
2. 麵粉、發粉、肉桂粉及豆蔻粉篩勻,備用。
3. 牛油與黃砂糖打透至忌廉狀,雞蛋逐個加入拂勻,加入已浸透的乾果拌勻,粉料分兩次拌入。
4. 預備 2 個 8 吋 x3 吋的長方形餅模,鋪上牛油紙及注入麵糊,放入預熱焗爐以 170℃ 焗約 20 分鐘,再轉 150℃ 焗 50 分鐘,待片刻取出。
5. 澆上冧酒,蓋上錫紙待片刻品嚐。

Method

1. Finely chop all dried fruits. Soak dried fruits and hazelnuts in rum overnight (or longer).
2. Sieve flour, baking powder, ground cinnamon and cardamom together. Set aside.
3. Beat butter and brown sugar until thick and pale. Add one egg after another and beat well after each addition. Put in the rum-infused dried fruits and nuts. Stir well. Pour in half of the dry ingredients from step (2). Mix well. Pour in the remaining half and mix again.
4. Prepare two 8" x 3" rectangular loaf tins. Pour the batter into the loaf tins lined with parchment paper. Bake in a preheated oven at 170°C for 20 minutes. Then turn down to 150°C and bake for 50 minutes. Leave the cakes in the oven briefly before taking them out.
5. Pour the rum over the cakes and cover the aluminium foil. Leave it a while. Serve.

家庭聚餐

每年，難得有幾個節日一家人共聚同歡，除了傳統的佳節之外，長輩的生日、父親節、母親節等等，是凝聚家人、增進感情的好時機。

準備餐前小食、主菜及甜品細意品嚐，全家人說說笑笑，暖意滿屋。以下的菜單特別選用了蟹肉、牛腱、白鱔及雞肉，配搭不同的醬汁炮製，每款都令人印象深刻，定會受到親友的讚賞。最後，來一道滋潤肺部的「杏汁燉木瓜」，幫助家人增強體質，祝願他們身體安康、萬事順心；一家人平安喜樂，是每個人的期盼。

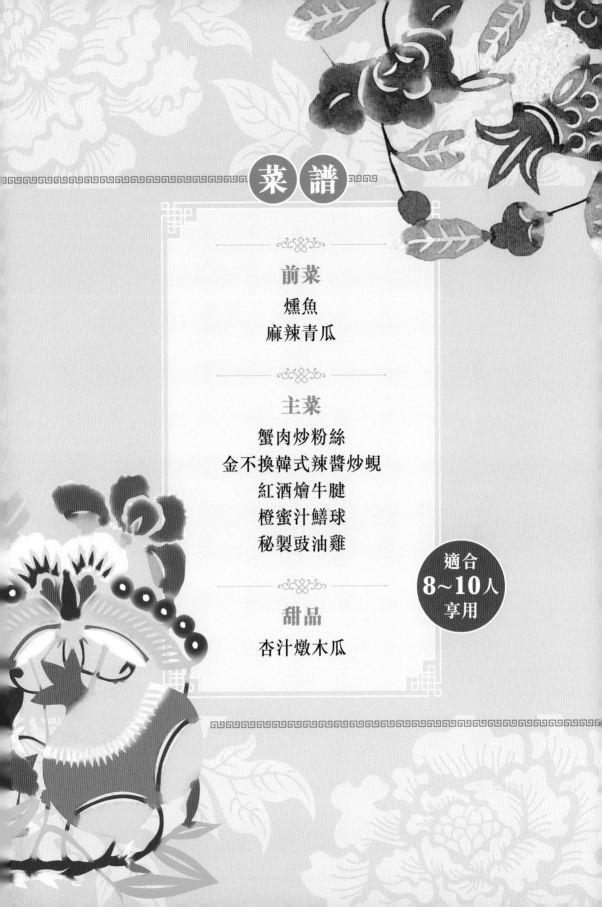

菜譜

前菜
燻魚
麻辣青瓜

主菜
蟹肉炒粉絲
金不換韓式辣醬炒蜆
紅酒燴牛腱
橙蜜汁鱔球
秘製豉油雞

適合
8~10人
享用

甜品
杏汁燉木瓜

燻魚

Shanghai Style Smoked Carp

【材料】

鯇魚腩 1.5 斤（切開約 8 至 9 塊）
粟粉 3 茶匙

【滷水汁料（蘇式浸汁）】

香葉 6 片
八角 5 粒
桂皮 1 塊
花椒 1 湯匙
老薑 2 塊（切片）

Ingredients

900 g grass carp (cut into 8-9 pieces)
3 tsp cornflour

Ingredients for Shanghai stock sauce

6 dried bay leaves
5 star anises
1 piece cinnamon
1 tbsp Sichuan peppercorns
2 pieces mature ginger (sliced)

【調味料】	【醃料】	Seasoning	Marinade
鹽 1 茶匙	鹽 1 茶匙	1 tsp salt	1 tsp salt
冰糖 2.5 湯匙	紹酒 1 湯匙	2.5 tbsp rock sugar	1 tbsp Shaoxing wine
老抽 1 湯匙		1 tbsp dark soy sauce	
紹酒 1 湯匙		1 tbsp Shaoxing wine	

【做法】

1. 滷水汁料與清水 4.5 杯煮滾，加入調味料用小火煮半小時，備用。
2. 鯇魚腩刮淨黑膜，洗淨，抹乾水分，切件，下醃料抹勻醃半小時。
3. 鯇魚腩加入粟粉拌勻，放入油鍋半煎炸至鯇魚腩呈金黃香脆，隔油。
4. 煮滾滷水汁，排入鯇魚腩，用小火煮滾 3 分鐘，熄火浸 10 分鐘，上碟，待冷食用。

Method

1. Boil together 4.5 cups of water and ingredients for the stock sauce. Add seasoning, turn to low heat and boil for 30 minutes. Set aside.
2. Scrape off any black tissue from grass carp. Rinse and wipe dry. Cut into pieces and mix well with marinade. Let it sit for 30 minutes.
3. Mix grass carp with cornflour. Fried in an oiled pan until browned and crispy. Drain off excessive oil.
4. Bring the stock sauce to boil. Add grass carp and boil over low heat for 3 minutes. Turn off heat and let it sit for 10 minutes. Transfer to a plate, let it cool and serve.

小技巧 Cooking tips

- 調味料加了冰糖，令燻魚的效果更晶瑩剔透。
- 煎至金黃香脆的鯇魚腩留有餘熱，能加快吸收滷水汁的香味，建議燻魚最多浸汁約 10 分鐘，以免味道太鹹。
- The rock sugar in the seasoning makes the grass carp glossy.
- Warm and fried grass carp can absorb the sauce very fast. It is recommended to soak for 10 minutes, it would be too salty if it is soaked for too long.

麻辣青瓜

Cold Cucumber Appetizer Dressed in Sichuan Peppercorn Chilli Oil

【材料】

溫室小青瓜半斤
雲耳 1/3 兩
蒜蓉 2 茶匙

【調味料】

麻香辣椒油 2 茶匙 *
麻油 2 茶匙
鹽、糖各半茶匙

* 麻香辣椒油

材料：川椒粒 4 湯匙、指天椒 4 兩（切碎）、豆豉 1 湯匙（切碎）、蝦米 2 湯匙（切碎）、乾葱蓉 2 湯匙、蒜蓉 1 湯匙、粟米油 1.5 杯

調味料：鹽及糖各 1 茶匙、生抽 1 湯匙

做法：燒熱油，下川椒粒用小火炸至香，隔去大部分川椒粒，下蝦米、指天椒及乾葱蓉，用小火炒至香，加入豆豉及蒜蓉炒勻，下調味料煮 5 分鐘，待涼。

【做法】

1. 雲耳用水浸軟，去硬蒂，洗淨，汆水，過冷河。
2. 小青瓜洗淨，切去頭尾兩端，用刀拍裂，再切成塊。
3. 將青瓜塊、雲耳、蒜蓉及調味料放容器內拌勻，冷藏片刻，作為前菜食用。

小技巧 Cooking tips

- 雲耳汆水煮透，可去除菇的霉味。
- 用刀拍裂的青瓜，咬入口質感佳，而且帶鬆脆口感。
- It's advisable to blanch the cloud ear fungus to cook it through, so as to remove its mouldy taste.
- Crushing the cucumber with the flat side of a knife makes it crack along its natural grain. The cucumber tends to have better mouthfeel with a lovely crunch.

Ingredients

300 g hothouse baby cucumbers
13 g cloud ear fungus
2 tsp grated garlic

Seasoning

2 tsp Sichuan peppercorn chilli oil*
2 tsp sesame oil
1/2 tsp salt
1/2 tsp sugar

Method

1. Soak the cloud ear in water until soft. Cut off the tough roots. Rinse well. Blanch in boiling water. Rinse with cold water.
2. Rinse the cucumbers. Cut off both ends and crush them with the flat side of a knife. Then cut into pieces.
3. Put the cucumber, cloud ear, garlic and seasoning into a container. Mix well. Refrigerate briefly. Serve.

* Sichuan peppercorn chilli oil

Ingredients: 4 tbsp Sichuan peppercorns, 150 g bird's eye chillies (finely chopped), 1 tbsp fermented black beans (finely chopped), 2 tbsp dried shrimps (finely chopped), 2 tbsp finely chopped shallot, 1 tbsp grated garlic, 1.5 cups corn oil

Seasoning: 1 tsp salt, 1 tsp sugar, 1 tbsp light soy sauce

Method: Heat the oil. Fry the Sichuan peppercorns over low heat until fragrant. Set aside most of the Sichun peppercorns. Add dried shrimps, bird's eye chillies and shallot. Stir fry over low heat until fragrant. Add fermented black beans and grated garlic. Add seasoning and cook for 5 minutes. Leave it to cool.

蟹肉炒粉絲

Stir-fried Crab Meat with Bean Vermicelli

【材料】

花蟹 1 隻（重約半斤）
粉絲 5 兩（大包裝計 1 包）
免治豬肉 5 兩
薑蓉 1 茶匙
葱粒 2 湯匙
紹酒 1/2 湯匙

【醃料】

生抽半湯匙
粟粉 1.5 茶匙
水 3 湯匙

【調味料】

生抽 1.5 湯匙
糖 2/3 茶匙

Ingredients

1 blue crab (about 300 g)
190 g bean vermicelli (1 pack from large pack)
190 g minced pork
1 tsp finely chopped ginger
2 tbsp diced spring onion
1/2 tbsp Shaoxing wine

Marinade

1/2 tbsp light soy sauce
1.5 tsp cornflour
3 tbsp water

Seasoning

1.5 tbsp light soy sauce
2/3 tsp sugar

【做法】

1. 花蟹洗淨外殼，隔水大火蒸 12 分鐘，待涼，去厴、去鰓及沙囊、拆肉備用。
2. 免治豬肉與醃料拌勻。
3. 粉絲用滾水浸 5 分鐘，過冷河，剪成短度，瀝乾水分。
4. 燒熱鑊下油 1 湯匙，下薑蓉炒香，加入蟹肉及蟹膏，潑酒炒勻，盛起。
5. 原鑊下油 2 湯匙，下免治豬肉炒至轉成白色，加入葱粒炒香，傾入熱水 2/3 杯及調味料煮滾，加入粉絲拌勻，下蟹肉炒至汁液收乾，散發陣陣蟹肉香味即成。

Method

1. Rinse the shell of the blue crab. Steam over high heat for 12 minutes. Leave to cool down. Remove the abdomen and gills. Take the meat and set aside.
2. Mix the minced pork well with the marinade together.
3. Soak the bean vermicelli in boiling water for 5 minutes. Rinse in cold water. Cut into short sections. Drain.
4. Heat up a wok. Put in 1 tbsp of oil. Stir-fry the ginger until aromatic. Add the crab meat and roe. Sprinkle with the wine and stir-fry evenly. Remove.
5. Add 2 tbsp of oil in the same wok. Stir-fry the minced pork until it turns white. Put in the spring onion and stir-fry until scented. Pour in 2/3 cup of hot water and the seasoning. Bring to the boil. Add the bean vermicelli and mix well. Put in the crab meat from step 4. Stir-fry until the sauce dries and it releases the fragrance of crab meat. Serve.

小技巧 Cooking tips

- 拆蟹肉時，只要拆掉蟹鰓、蟹厴及沙囊，再細心地用竹籤挑出蟹肉，其實並不困難。半斤重的花蟹得大半碗蟹肉及蟹膏，鮮甜無比！

- When you take the meat from the crab, just get rid of the crab's gills and abdomen and then pick out the meat with a bamboo skewer. A blue crab weighting 300 g can yield more than half bowl of meat and roe. Fresh crab meat gives an incomparably sweet flavour!

家庭聚餐

金不換韓式辣醬炒蜆

Stir-fried Clams with Thai Basil and Korean Chilli Sauce

【材料】
活蜆 1.5 斤
金不換 3 棵
紅辣椒 2 隻（切碎）
蒜肉 5 粒（拍鬆）
韓式辣椒醬 2.5 湯匙
紹酒 1 湯匙

【調味料】
魚露 1 茶匙

Ingredients

900 g live clams
3 stalks Thai basil
2 red chillies (finely chopped)
5 cloves garlic (bashed)
2.5 tbsp Korean chilli sauce
1 tbsp Shaoxing wine

Seasoning

1 tsp fish sauce

【做法】

1. 金不換摘葉，洗淨後備用。
2. 活蜆放在水中，擦洗多次，瀝乾水分。
3. 煮滾半鑊水，放入蜆灼至半開殼，盛起，瀝去水分。
4. 燒熱鑊下油 3 湯匙，下蒜肉炒香，加入蜆，潷酒炒勻，加入韓式辣椒醬、調味料、紅辣椒及熱水 3 湯匙炒勻，加蓋焗 3 分鐘，下金不換葉炒勻即成。

Method

1. Pick off the leaves of the basil and rinse. Set aside.
2. Put the live clams in water, rub and wash for many times and drain.
3. Bring half wok of water to the boil. Blanch the clams until half open, dish up and drain.
4. Heat a wok, put in 3 tbsp of oil, stir-fry the garlic until fragrant. Add the clams, sprinkle with the wine and stir-fry evenly. Add the Korean chilli sauce, seasoning, red chillies and 3 tbsp of hot water and then give a good stir-fry. Cover the wok and leave for 3 minutes. Finally add the basil and stir-fry evenly. Serve.

小技巧 Cooking tips

- 將蜆放入熱水略浸至半張開，可吐出砂粒，且將沒張開口的死蜆拿掉。
- Slightly soak the clams in hot water until they are half open to spill the sand. The closed clams, which are dead, can also be picked out.

紅酒燴牛䐆
Red Wine Stewed Beef Shin

【材料】
急凍牛䐆 1 斤
紅酒 1.5 杯
甘筍 1 條（約 8 兩）
洋葱半個
番茄 3 個
西芹 3 條
蒜肉 3 粒
迷迭香 2 條

【調味料】
黑椒碎 2 茶匙
鹽 1 茶匙
黃砂糖 1 茶匙
茄汁 2 湯匙

Ingredients

600 g frozen beef shin
1.5 cup red wine
1 carrot (about 300 g)
1/2 onion
3 tomatoes
3 pieces celery
3 cloves garlic
2 sprigs rosemary

Seasoning

2 tsp chopped black pepper
1 tsp salt
1 tsp brown sugar
2 tbsp ketchup

【做法】

1. 急凍牛腱解凍，洗淨，汆水，過冷河，切厚塊備用。
2. 甘筍去皮，洗淨、切塊；番茄去蒂，洗淨，切角；西芹洗淨，切塊；洋蔥去外衣，洗淨，切碎。
3. 燒熱鑊下油 3 湯匙，放入蒜肉、洋蔥炒香，加入甘筍、番茄、西芹炒勻，下牛腱拌勻，全部材料轉放鍋內，傾入紅酒、清水浸過牛腱表面，最後加入迷迭香煮滾，轉小火燜約 1 小時，下調味料再煮片刻，待牛腱軟腍即成。

Method

1. Defrost beef shin and rinse. Boil briefly and rinse again with cold water. Cut into thick pieces. Set aside.
2. Peel carrot, rinse and cut into pieces; remove stalks from tomatoes, rinse and cut into wedges; rinse celery and cut into pieces; peel onion, rinse and chop.
3. Heat wok and add 3 tbsp of oil. Fry garlic and onion until fragrant. Add carrot, tomatoes and celery and stir fry well. Add beef shin and stir well. Transfer in a pot. Add red wine and water until the beef shin is soaked under. Add rosemary and bring to boil. Turn to low heat and simmer for 1 hour. Add seasoning and boil for a while, until beef shin softens. Serve.

小技巧 Cooking tips

- 喝不完的紅酒可以用於烹調，紅酒很耐存，只要用酒塞封好，存放於陰涼的地方，任何時候烹調也可。
- You can use the leftover red wine for cooking. As long as the bottle is sealed and stored in shaded place, you can use it to cook.

橙蜜汁鱔球

Deep-fried Japanese Eel in Orange Honey Sauce

【材料】
白鱔 1/2 條（魚販代起骨）
粟粉 3 湯匙

【醃料】
鹽 2/3 茶匙
胡椒粉少許

【調味料】
橙汁 2/3 杯
蜜糖 3 湯匙
黃砂糖 1 茶匙
鹽 1/8 茶匙
粟粉 1 茶匙
水 2 湯匙

Ingredients

1/2 Japanese eel
(boned by the fishmonger)
3 tbsp cornflour

Marinade

2/3 tsp salt
ground white pepper

Seasoning

2/3 cup orange juice
3 tbsp honey
1 tsp brown sugar
1/8 tsp salt
1 tsp cornflour
2 tbsp water

【做法】

1. 白鱔肉放入熱水內浸半分鐘，用刀刮淨表面黏液，洗淨。
2. 用刀在白鱔肉的一面，斜剉成格子紋狀，切塊，下醃料拌勻待半小時。
3. 白鱔與粟粉拌勻，放入滾油內用中小火炸至捲起及金黃全熟，盛起隔油。
4. 調味料用慢火煮成糊狀，拌入鱔球，上碟享用。

Method

1. Soak the Japanese eel in hot water for half a minute. Scrape mucus off the skin with a knife and then rinse.
2. Score one side of the Japanese eel diagonally in crisscross patterns and then cut into pieces. Mix with the marinade and rest for half an hour.
3. Mix the Japanese eel and cornflour together. Put into boiling oil and deep-fry over medium-low heat until it curls up and is brown and cooked through. Drain.
4. Simmer the seasoning sauce into a paste. Mix in the Japanese eel. Put on a plate and serve.

小技巧 Cooking tips

- 用刀剉成格子紋時，別切得太貼近魚皮，以免魚肉容易散開，影響賣相。

- When scoring the meat of Japanese eel in a crisscross pattern, do not go deep down to the skin; otherwise, the meat will fall apart affecting the presentation.

秘製豉油雞

Soy Sauce Chicken

【材料】

冰鮮雞 1 隻（約重 2 斤）
薑 6 片
蔥 3 條
八角 2 粒
冰糖 1 湯匙

【豉油汁】

老抽 3 湯匙
生抽 2 湯匙
紹酒 1 湯匙
水 3 杯

Ingredients

1 chilled chicken (about 1.2 kg)
6 slices ginger
3 sprigs spring onion
2 star aniseed
1 tbsp rock sugar

Mixed soy sauce

3 tbsp dark soy sauce
2 tbsp light soy sauce
1 tbsp Shaoxing wine
3 cups water

小技巧 Cooking tips

- 自製的豉油汁料非常簡單，包括薑蔥、八角、冰糖、老抽、生抽及紹酒，由於冰糖容易令雞皮焦燶，宜用易潔鑊烹調，若使用不銹鋼鍋應先鋪上竹笪。

- 為了雞皮均勻上色，翻動雞後用小火加蓋每邊浸 5 分鐘即可。

- The ingredients for making soy sauce by ourselves are very simple, namely ginger, spring onion, star aniseed, rock sugar, dark soy sauce, light soy sauce and Shaoxing wine. It is better to use a non-stick pan to cook because the rock sugar will make the chicken skin burnt easily. If a stainless steel pot is used, place a bamboo mat on it first.

- Turn the chicken and simmer each side with a lid on for 5 minutes to colour the skin easily and evenly.

【做法】

1. 冰鮮雞去掉肺部,用粗鹽擦洗雞尾部分,洗淨,瀝乾水分。
2. 葱去掉鬚根,洗淨。
3. 豉油汁放入鍋內,加入薑、葱、八角及冰糖煮滾,放入雞加蓋,用中火煮 15 分鐘,翻動雞隻,調至小火加蓋,每邊浸 5 分鐘至熟透,待涼,斬件上碟,澆上豉油汁享用。

Method

1. Remove the lungs of the chicken, rub the buttock with coarse salt, rinse and drain.
2. Cut away the root of the spring onion and then rinse.
3. Put the mixed soy sauce into a pot, add the ginger, spring onion, star aniseed and rock sugar, and then bring to the boil. Put in the chicken, put a lid on and cook over medium heat for 15 minutes. Turn over the chicken, adjust to low heat and put the lid on. Soak each side for 5 minutes or until it is cooked through. When it cools, chop into pieces, drizzle with the soy sauce. Serve.

杏汁燉木瓜

Double-steamed Papaya in Almond Milk

【材料】

南杏仁 4 兩

北杏仁 2 湯匙

半生熟木瓜 1 個

冰糖 2 湯匙（舂碎）

Ingredients

150 g sweet almonds

2 tbsp bitter almonds

1 half-ripe papaya

2 tbsp rock sugar (crushed)

一家團圓節慶菜

【做法】

1. 南北杏洗淨，用水浸 1 小時，隔去水分。
2. 南北杏及清水 1.5 杯放於攪拌機內，磨成杏仁漿，用隔篩過濾，即成幼滑的杏仁汁。
3. 木瓜去皮、去籽，切塊。
4. 煮滾清水 4 杯，加入冰糖及杏汁拌勻，煮至微滾，將杏汁傾入燉盅內，加入木瓜肉，加蓋，隔水中火燉半小時即可。

Method

1. Rinse the sweet and bitter almonds. Soak in water for 1 hour. Strain.
2. Put all the almonds and 1.5 cup of water into a blender. Grind into almond milk. Sieve out the creamy almond milk.
3. Peel the papaya. Remove the seeds and cut into pieces.
4. Bring 4 cups of water to the boil. Add the rock sugar and almond milk. Mix well. Cook until it slightly boils. Pour the almond milk into a covered ceramic container. Add the papaya. Cover with the lid. Double-steam over medium heat for 1/2 hour. Serve.

小技巧 Cooking tips

- 杏汁煮至微滾後放入燉盅，令杏汁快速達至溫點，縮短燉煮的時間。
- The almond milk is slightly boiled before transferring to a covered ceramic container. It is to heat up the almond milk quickly, shortening the time for double-steaming.

編著者
萬里編輯委員會

責任編輯
簡詠怡

裝幀設計
鍾啟善

排版
楊詠雯、劉葉青

出版者
萬里機構出版有限公司
香港北角英皇道 499 號北角工業大廈 20 樓
電話：(852) 2564 7511
傳真：(852) 2565 5539
電郵：info@wanlibk.com
網址：http://www.wanlibk.com
　　　http://www.facebook.com/wanlibk

發行者
香港聯合書刊物流有限公司
香港荃灣德士古道 220-248 號荃灣工業中心 16 樓
電話：(852) 2150 2100
傳真：(852) 2407 3062
電郵：info@suplogistics.com.hk
網址：http://www.suplogistics.com.hk

承印者
美雅印刷製本有限公司
香港九龍觀塘榮業街 6 號海濱工業大廈 4 樓 A 室

規格
特 16 開（240mm x 170mm）

出版日期
二〇二〇年十二月第一次印刷

6 個中西節日 ✕ 48 道美味佳餚